InSTREAM: The Individual-Based Stream Trout Research and Environmental Assessment Model

Steven F. Railsback, Bret C. Harvey, Stephen K. Jackson, and Roland H. Lamberson

U.S. Department of Agriculture, Forest Service

Pacific Southwest Research Station

Albany, California

General Technical Report PSW-GTR-218

August 2009

Published in cooperation with:

U.S. Environmental Protection Agency

National Center for Environmental Research STAR Program

Abstract

Railsback, Steven F.; Harvey, Bret C.; Jackson, Stephen K.; Lamberson, Roland H. 2009. InSTREAM: the individual-based stream trout research and environmental assessment model. Gen. Tech. Rep. PSW-GTR-218. Albany, CA: U.S. Department of Agriculture, Forest Service, Pacific Southwest Research Station. 254 p.

This report documents Version 4.2 of InSTREAM, including its formulation, software, and application to research and management problems. InSTREAM is a simulation model designed to understand how stream and river salmonid populations respond to habitat alteration, including altered flow, temperature, and turbidity regimes and changes in channel morphology. The model represents individual fish at a daily time step, with population responses emerging from how individuals are affected by their habitat and by each other (especially, via competition for food). Key individual behaviors include habitat selection (movement to the best available foraging location), feeding and growth, mortality, and spawning. Fish growth depends on prey availability and hydraulic conditions. Mortality risks due to terrestrial predators, piscivorous fish, and extreme conditions are functions of habitat and fish variables. Field and analysis techniques for applying InSTREAM are based in part on extensive analysis of the model's sensitivities and uncertainties. The model's software provides graphical displays to observe fish behavior, detailed output files, and a tool to automate simulation experiments.

Keywords: Individual-based model, instream flow, population model, water temperature, Salmonidae, stream, trout, turbidity.

Summary

This report documents Version 4.2 of InSTREAM, including its formulation, software, and application to research and management problems. InSTREAM is a simulation model designed to understand how stream and river salmonid populations respond to habitat alteration, especially altered flow, temperature, and turbidity regimes. This model has been used at sites ranging from headwaters to large regulated rivers for a variety of research and management questions. It is a complex model with many inputs and parameters, but because InSTREAM directly predicts the cumulative effect of multiple variables on fish populations, it does not require the kinds of untested assumptions and guesswork needed to base decisions on simpler models.

Chapter 1 provides guidance on determining whether InSTREAM is a good tool for a particular situation.

Chapter 2 describes InSTREAM's formulation: its detailed assumptions and parameters and the information they were based upon. The model represents individual fish at a daily time step, with population responses emerging from how individuals are affected by their habitat and by each other (especially, via competition for food). Habitat is represented as microhabitat cells within one or several stream reaches. Key individual behaviors are habitat selection (movement to the best available foraging location), feeding and growth, mortality, and spawning. Fish growth depends on prey availability and hydraulic conditions. Mortality risks owing to terrestrial predators, piscivorous fish, and extreme conditions are functions of habitat and fish variables. Spawning and redd incubation are modeled simply while still representing effects of flow and temperature on reproductive success. Because the full life cycle is simulated, InSTREAM is typically used to simulate populations over one or more decades.

Chapter 3 describes field and analysis techniques for applying InSTREAM. These techniques are based on the authors' experience and on extensive analysis of the model's sensitivities and uncertainties.

Chapter 4 is an extensive guide to the model's software. The software provides graphical displays to observe fish behavior, detailed output files, and a tool to automate simulation experiments.

Contents

Chapter 1: Introduction and Overview

1. Introduction, Objectives, and Background

1.1. Document Purpose and Overview

This document describes the Individual-based Stream Trout Research and Assessment Model: why InSTREAM was developed, how it was designed and implemented, and how it can be used for environmental impact assessment and research. InSTREAM Version 4.2 is the first version intended for public use. Therefore, this report is intended to help new users decide whether to use InSTREAM, understand what the model does, and design and carry out applications.

The report generally follows the protocol recommended by Grimm et al. (2006) for describing individual-based models, although the report addresses topics (in chapters 3 and 4) not addressed by Grimm et al.

Chapter 1 is intended to help new users develop an understanding of InSTREAM and its applications. It summarizes the history of InSTREAM and its predecessors, then describes the model's basic design philosophy and fundamental assumptions. A summary of the environmental and biological processes represented in the model is then provided. Finally, the terminology and conventions used in InSTREAM are listed. Readers are strongly encouraged to become familiar with the terminology from the start, as a number of common words (e.g., action, method, object) have very specific and important meaning in this document.

Chapter 2 provides a detailed description of the model's assumptions, including equations, parameters, and inputs. It documents **why** these assumptions were chosen, and how they have been calibrated and tested.

Chapter 3 is intended to help users adapt applications of InSTREAM to new sites and study problems. It includes guidance on designing model-based studies, collecting field data for input, calibrating the model, and evaluating uncertainties in results.

Chapter 4 is a users' guide to InSTREAM's software. This guide covers installing the model and the Swarm simulation system that it runs in, preparing input files, using output files and graphical output, and preparing automated experiments with multiple scenarios and replicates. It also provides guidance for making some basic, easy changes to the software, for example to change how many or which trout species are simulated, add or remove mortality risks, and modify the statistical summary output files.

Users and potential users of InSTREAM are encouraged to periodically check the Web site for individual-based ecological modeling at Humboldt State University: http://www.humboldt.edu/~ecomodel for improvements and fixes to InSTREAM, its software, and this documentation.

1.2. Objectives of InSTREAM

As its name indicates, InSTREAM can support both environmental impact assessment and ecological research. The model's original purpose was to address one of the most difficult general problems of impact assessment for stream-dwelling trout: understanding how alteration of habitat affects populations of animals that actively adapt to habitat change by moving. InSTREAM can predict how trout populations respond to changes in any of the inputs that drive the model, especially flow, temperature, turbidity, and channel morphology. InSTREAM can also predict how populations respond to changes in ecological conditions such as food availability or mortality risk. Because InSTREAM provides an observable virtual ecosystem, it is also a useful tool for addressing many basic ecological research questions.

The complexity of InSTREAM, which no doubt is a concern to many potential users, is determined by the model's objectives. As we designed InSTREAM, we strove to keep it as simple as possible. However, it was clear that meeting its objectives requires modeling how population dynamics emerge from trout habitat and behavior, which requires explicitly representing such processes as feeding, bioenergetics, different causes of mortality, and the spawning and incubation cycle. The need to include these processes means InSTREAM has many equations and parameters. Traditional simple ecological models generally do not describe key processes mechanistically, so their parameters can only be estimated by fitting the model to data (e.g., Hilborn and Mangel 1997). Thus, many modelers assume complex models are inherently uncertain because their many parameters must be fit to data. This generalization does not hold for models such as InSTREAM, which can be viewed as a collection of submodels for key processes: the equations and parameters of each submodel are supported independently, often by extensive literature. Very few of InSTREAM's parameters are considered "free" for calibration. Sections 1.4, 15, and 17 provide additional discussion of uncertainty and parameter fitting.

Environmental assessment applications—
The primary motivation for developing InSTREAM was to provide an improved instream flow assessment model for comparing the fish population benefits of alternative streamflow regimes. InSTREAM was designed to overcome many of the limitations of PHABSIM (Bovee et al. 1998), the instream flow model now most widely used. The most important of these limitations are that PHABSIM (1) is static and cannot evaluate the effects of variability in flow (e.g., changing the timing of low vs. high flows); (2) predicts only changes in habitat, not population responses; (3) relies on the questionable assumption that habitat "preference" (the degree to

which a habitat type is selected by fish) indicates habitat quality; and (4) neglects how preference varies with many habitat and biological factors (Railsback 1999a). InSTREAM avoids all of these limitations, and, in fact, research using InSTREAM has shown that use of PHABSIM can lead to erroneous conclusions about how trout populations respond to changes in flow (Railsback et al. 2003).

A second purpose of InSTREAM is to assess effects of variations in stream temperature on fish populations. Management decisions that affect flow generally affect temperature as well. InSTREAM was designed to simulate how temperature affects trout both directly by mortality of fish or eggs at extreme temperatures and indirectly by changes in behavior and bioenergetics.

Assessing the effects of turbidity is another management purpose of InSTREAM. Sediment loading and turbidity are management concerns in many watersheds. Turbidity can have both positive and negative effects: increasing turbidity reduces the risk of predation on trout but reduces their ability to feed. InSTREAM was designed to predict the population-level consequences of these separate effects, a role that no previous models have filled.

Channel modification and restoration is another management issue the model can address. More attention is being paid to geomorphic consequences of watershed and river management decisions: How do dams, timber harvest, and land use changes affect geomorphic processes and, over the long term, alter river channels (e.g., Ziemer et al. 1991)? What are the effects of changes in channel morphology and sediment size on fish populations? At the same time, expensive channel restoration projects are increasingly proposed or conducted to improve fish habitat (Kondolf 2000). Predicting or assessing the benefits of restoration projects has therefore become an important management problem that InSTREAM can address. Changes in channel morphology, hydraulics, and the availability of cover for feeding and hiding can be input to InSTREAM, and the model will predict how the changes affect trout populations.

Finally, one of the most important among the unique characteristics of InSTREAM as an assessment tool is that it can address cumulative effects of multiple stressors (Harvey and Railsback 2007). The need for instream flow assessment methods that consider such processes as population dynamics, energetics, competition, and predation risk has long been recognized (e.g., Orth 1987). InSTREAM represents how population dynamics emerge from growth, survival, and reproduction of individuals, whether these processes are affected by flow, temperature, turbidity, or combinations of stressors. Other stressors with well-understood effects on individuals, such as pollutants and angler harvest, could also be added to InSTREAM. As long as the effect of a stressor on individuals can be quantified, InSTREAM can predict how the stressor interacts with others to affect the simulated trout population.

Research applications—

Several research applications of inSTREAM are mentioned above, and many more are possible. One additional research application is understanding relations among habitat, behavior, and population responses. To establish the validity of InSTREAM's approach for modeling how fish move among microhabitats, Railsback and Harvey (2002) used the model to compare alternative theories for habitat selection. Harvey and Railsback (2009) used the model to develop general understanding of how increased turbidity, mediated by habitat selection behavior, affects long-term population dynamics and persistence. One of the questions addressed by Railsback et al. (2002) is how channel shape (the availability of deep pools) affects population age and size structure. Another question addressed by Railsback et al. (2002) is density dependence of the size of juvenile trout. Observed patterns of density dependence were reproduced in InSTREAM, but model results were inconsistent with the presumed causal mechanism (density effects on food intake and growth); size-dependent mortality was indicated as an alternative explanation. Potential climate change applications include understanding how changes in temperature, natural flows, and increased water diversion affect population persistence (Xenopoulos et al. 2005).

A second general kind of research application is using InSTREAM as a virtual ecosystem in which to test ecological theory. Railsback et al. (2002), for example, used InSTREAM to address a classic problem of ecological theory: whether, and why, populations follow a "-4/3 power law" self-thinning relationship. This study provided a unique contribution to this topic because experiments that are impossible in nature (e.g., manipulating the animals' respiration allometry) are easy in InSTREAM. Similarly, Railsback et al. (2003) used InSTREAM to address a critical question that is virtually impossible to study in nature: What is the relationship between the density of animals using a habitat and the habitat's fitness value to an individual?

1.3. The Evolution of InSTREAM

The fundamental assumptions and detailed design of InSTREAM resulted from the merger of two long pathways of model evolution. The first pathway is the development of individual-based models (IBMs) as a way to solve specific problems in assessment and management of electric power production effects on fisheries. This pathway can be traced as far back as the decades-long regulatory battles over the importance of entrainment mortality of larval fish at thermal-electric powerplants in the Hudson River estuary (Barnthouse et al. 1988). One outcome of this battle was the realization that conventional ecological models cannot address the question of how density-dependence of survival of juvenile fish might compensate for the entrainment losses. The IBMs were identified as a type of model that could address this question, and the Electric Power Research Institute (EPRI) began sponsoring research on fish IBMs through its Compensation Mechanisms in Fish Populations (CompMech) program. Eventually, EPRI-funded researchers at Oak Ridge National Laboratory identified instream flow assessment as another important problem for which IBMs have many potential advantages. The first EPRI IBM for stream fish simulated smallmouth bass (*Micropterus dolomieu*) (Jager et al. 1993). The second was a trout IBM (Van Winkle et al. 1996, 1998) that became the predecessor to InSTREAM.

The second pathway in the evolution of modeling from which InSTREAM descends is the so-called complexity movement, which has its intellectual home at the Santa Fe Institute in New Mexico (Holland 1995, Levin 1999, Waldrop 1992). One outcome of complexity science is a set of biology-based concepts—emergence, adaptation, fitness, etc.—that provide a useful framework for designing and describing IBMs (Railsback 2001). These concepts are used throughout this report. A second outcome that InSTREAM takes very direct advantage of is the Swarm simulation system for agent-based models and IBMs, originally developed by Chris Langton and others at the Santa Fe Institute (Minar et al. 1996). Not only is InSTREAM implemented using the Swarm modeling framework and software, the community of Swarm developers and users have contributed in many ways to the model's design.

These two pathways came together at Humboldt State University in 1998, when Pacific Gas & Electric Company and Southern California Edison Company funded a complete redesign and reimplementation of the CompMech two-species trout IBM (referred to now as version 1 of InSTREAM). Subsequent work funded by EPRI and the USDA Forest Service made many refinements and tests of both the model's formulation and its software (Railsback 1999b, Railsback and Harvey 2001). Versions 2 and 2.2 implemented such changes as adding turbidity as a driving variable; simulating multiple, linked, stream reaches; and adding software for automated simulation experiments. Version 3, developed for EPRI and Western Area Power Administration, considers within-day flow changes, how feeding success and predation risks differ between day and night, and how trout choose whether to feed vs. hide during day and night (Railsback et al. 2005).

Starting in 2003, the U.S. Environmental Protection Agency (USEPA) has funded development and analysis of InSTREAM as a watershed-scale decision-support tool. This support is provided through USEPA's Science to Achieve Results program 2002 solicitation "Developing Regional-Scale Stressor-Response Models for Use in Environmental Decisionmaking." As a result of this funding, the version 4 series, including the version 4.2 documented in this report, are the first designed specifically for public distribution and use for management applications.

As part of the studies described above, InSTREAM has been applied to a number of study sites. The primary sites used for model development and testing continue to be three reaches of Little Jones and Weejak Creeks, which are third- and first-order streams in the Smith River basin, Del Norte County, California (Harvey 1998, Railsback and Harvey 2001). A reach of upper Jacoby Creek near Arcata, Humboldt County, California, has also been modeled. Version 3 has been applied to four reaches of the large Green River below Flaming Gorge Dam, Daggett County, Utah. Starting in 2004, four new study sites were established in the Eel River watershed in Humboldt and Mendocino Counties, California. The latest version of InSTREAM (4.3; documented separately) uses a two-dimensional hydrodynamic model to simulate cell boundaries, depths, and velocities; it is designed for larger streams and rivers.

1.4. Is InSTREAM the Right Tool?

Perhaps the first question a potential user of InSTREAM must address is whether the model is an appropriate tool for some particular river management or research problem. Section 1.2 lists some specific kinds of assessment and research problems that InSTREAM was designed to address. In general, InSTREAM was designed for studies of how changes in physical habitat (channel morphology, flow, temperature, and turbidity) affect the long-term production and persistence of trout populations. InSTREAM has also proven useful for studies of how trout populations are affected by biological processes that are represented in the model. Examples include examining interactions and competition among multiple trout species (or even among age classes of the same species) and effects of changes in food production or predation risk.

There are no clear limits to the physical characteristics of sites (size, slope, geomorphology, etc.) where InSTREAM could be applied (but see the following discussion of sites where the model would be inappropriate). As discussed in chapter 2, a few of the model's detailed assumptions are clearly more valid for particular kinds of streams. An example is the method for modeling mortality of trout eggs owing to sediment scouring or deposition, which was designed for gravel-bedded streams.

For any application, one should review model details and determine whether any processes should be turned off (via parameter values) or altered.

Clearly, InSTREAM is not appropriate for study sites or problems where the model's fundamental assumptions are not met or where trout population dynamics are strongly dependent on processes that are not adequately represented in the model. Some examples of sites or problems where InSTREAM may not be appropriate (without modification) are:

- Sites where nonsalmonid species are significant competitors for food or habitat.
- Sites where water quality constituents other than temperature and turbidity have strong effects or are the management issues of interest. Dissolved oxygen, for example, is not considered in InSTREAM. Likewise, the effects of fine sediment on egg incubation are not considered.
- Study problems involving adaptive behaviors of trout that are not in InSTREAM. For example, the model does not represent how life history characteristics adapt via behavior or evolution, so it is not useful for addressing problems of life history adaptation.

Of special concern are sites where the effects of ice are important. Ice can cause direct mortality, alter or exclude habitat, reduce invertebrate food production via scouring, and provide protection from predation. None of these processes are now included in InSTREAM, in part because they are difficult to model; even the presence of ice is difficult to predict. Potential users of InSTREAM would have to consider whether ice effects can be neglected, or how they could be represented in the model in a simple way. Direct mortality from ice could be coarsely represented as an additional survival probability (section 6.4) that varies with water temperature, date, etc.; ice effects on invertebrates could be represented coarsely by assuming food availability varies with the same kinds of conditions. Ice effects should be less important where management actions do not affect ice (e.g., changes in turbidity or summer flow), so the ice effects are constant among alternatives. But InSTREAM would require significant modification to provide a mechanistic representation of how trout populations respond to management actions that strongly affect ice (e.g., altering winter flow or temperature).

The complexity of InSTREAM is also a concern in deciding whether to use it, especially in contentious situations where methods and results will be heavily scrutinized. There is a widespread belief that models such as InSTREAM that have many assumptions and parameters are inherently more uncertain than simpler models and more subject to bias. Certainly, the prospect of having to justify all of InSTREAM's equations and parameters in litigation would be a disincentive to use it instead of a more established approach such as PHABSIM. However, there are reasons to doubt that InSTREAM is inherently more uncertain or subject to bias than more established models. First, InSTREAM explicitly represents important processes (described in section 1.2) that are ignored in models such as PHABSIM; when ignored, these processes must instead be dealt with ad hoc as decisions are made, which adds uncertainty. Dealing with results for multiple life stages is a perfect example: PHABSIM produces separate results for spawning, juveniles, and adults; and there are no well-justified methods for combining these results into meaningful population-level results. InSTREAM directly predicts population status from what happens at every life stage. Second, habitat preference models such as PHABSIM depend almost entirely on preference curves, and these curves are essentially a large parameter set. Hence, these models also have many parameters. Third, this document defines and justifies each equation and parameter value. Finally, the robustness analyses described in chapter 3 show that InSTREAM's results for decisionmaking—the ranking of management alternatives by the trout populations they provide—can be quite robust to parameter uncertainty.

2. Overview of InSTREAM

2.1. Modeling Philosophy

The design of InSTREAM was guided by a few general modeling principles that users should be aware of. First, and most important for all models, is that the model's design was strongly determined by its purposes and the drive for simplicity. Variables, processes, and assumptions were included only if they appeared necessary to meet the model's purpose of predicting trout population response to changes in habitat conditions. Some habitat and trout characteristics and behaviors that may seem important to biologists were excluded from the model because they would add complexity that makes the model harder to test and understand without being essential for meeting its purpose.

The model includes a number of processes and variables that are clearly important but not well understood because of their natural complexity; food availability and predation mortality are perhaps the best examples. For such elements of the model, the approach was to use the simplest possible assumptions that appeared useful and reasonable. However, in a few cases (especially, the bioenergetics equations for modeling growth) more complex approaches were used because credible equations and parameters were available and well-established in the literature.

The philosophy of keeping InSTREAM as simple as possible extends to how different trout species are represented. Trout biologists will note that the model's design and our guidance for representing multiple species (chapter 3) ignore many of the ways that trout species are often believed to differ from each other. One reason for minimizing built-in differences among species is to keep the model simple; another is to avoid "hardwiring" apparent differences among species (or populations of the same species) that may instead emerge simply from differences in site conditions.

Another key modeling principle is that all parts of InSTREAM must be observable and testable. As many of the model's assumptions and methods as possible were tested independently as they were designed (documented in chapter 2). The primary adaptive behavior of trout in InSTREAM is habitat selection. The habitat selection approach in InStream could not be tested independently, therefore we validated it by demonstrating the model's ability to reproduce a variety of habitat selection patterns observed in real trout (Railsback and Harvey 2002). Providing the ability to observe trout, their habitat, and their behavior was a key objective in designing the software for InSTREAM (chapter 4).

2.2. Fundamental Assumptions

The first step in understanding and using InSTREAM is to understand its fundamental assumptions. These assumptions are presented using the conceptual framework for IBMs suggested by Railsback (2001) and Grimm and Railsback (2005).

Emergence—
The most fundamental assumption of IBMs such as InSTREAM is that population responses **emerge** from processes acting at the individual level. In InSTREAM, population responses include many characteristics of real fish populations: abundance; biomass; production; statistical distributions of age, weight, and length; habitat use patterns; and measure of persistence such as mean time to extinction. These population characteristics emerge from the growth, survival, and reproduction of individuals, whereas these individual-level processes are affected by habitat inputs such as flow, temperature, turbidity, and channel shape.

Adaptation—
How population responses emerge from individual growth, survival, and reproduction is strongly determined by how the individuals adapt to changes in themselves and their habitat. In InSTREAM, the primary adaptive trait used by trout is **habitat selection** (also called **movement**): their decision of which habitat cell to occupy each time step. (The term "habitat selection" is also used for models based on the assumption that animals have innate "preferences" for certain ranges of habitat variables such as depth and velocity (e.g., Manly et al. 2002). Although both kinds of model predict the choice of microhabitat, the habitat selection trait in

InSTREAM is not based on preferences.) Other adaptive traits are selecting which of two feeding strategies a fish uses each day; and the decision by adult females of when and in which cell to spawn.

Trout have additional adaptive behaviors that we have chosen **not** to represent mechanistically in this version of InSTREAM, because doing so does not seem necessary to meet the model's purposes. These behaviors include variation in diel activity patterns (feeding vs. hiding); allocation of energy intake to growth, energy storage, or gonad production; year-to-year spawning effort.

Fitness—

The habitat selection trait is modeled as a fitness-seeking process in which trout select the cell that offers the highest value of a measure of expected fitness. The fitness measure used in InSTREAM is the "Expected Reproductive Maturity" measure developed by Railsback et al. (1999) and tested by Railsback and Harvey (2002), described in section 6.2.

Interaction—

Trout compete for food and feeding habitat (velocity shelters) according to a length-based hierarchy. Each habitat cell contains a limited daily food supply and a fixed area of velocity shelter. Food consumption by larger individuals potentially limits the amount of food a trout could get if it occupied the same cell. Similarly, larger fish reduce the area of velocity shelter available.

Stochasticity—

InSTREAM is not a highly stochastic model. Mortality is the most important process represented stochastically. The daily **probability of survival** for each trout is a deterministic function of its state and its habitat; but whether the trout actually lives or dies each day is a stochastic event. The other use of stochasticity is initializing a model run: input files specify how many trout of each age are to be initialized, and the mean and standard deviation of length for each age class. The actual length of each individual is drawn from a random distribution with the specified mean and standard deviation; its sex is assigned randomly; and the individual's initial location is selected randomly from the cells with nonextreme depths and velocities. A similar approach is used to assign the length of new fish produced in the model as they hatch from eggs. Another stochastic process determines whether a female that is ready to spawn actually does spawn on a particular day. Some of the methods for representing mortality of incubating eggs are also partially random. The model incorporates environmental variability through the driving physical variables: stream discharge, turbidity, and temperature.

Spatial scales—

Space is represented as a collection of discrete, nonuniform, rectangular cells (fig. 1); habitat conditions differ among cells but not within a cell. The spatial resolution is therefore the size of one cell. Cell dimensions are variable and chosen in the field to represent the actual variation in microhabitat variables such as depth, velocity, and cover availability. Inoue and Nunokawa (2002), for example, found that habitat selection could be relatively well explained at this sub-habitat-unit scale. The area typically occupied by one adult trout (very roughly, 1 m^2) is used as a lower limit on cell size. The spatial extent (the total area simulated) is chosen by the user as a tradeoff between representing the study site (better with larger areas) and the field data and computational effort needed to represent larger areas. The spatial extent of InSTREAM can include multiple, linked stream reaches. There are no restrictions on how many cells can be in a reach, how many reaches can be in a model, or how multiple reaches are arranged spatially.

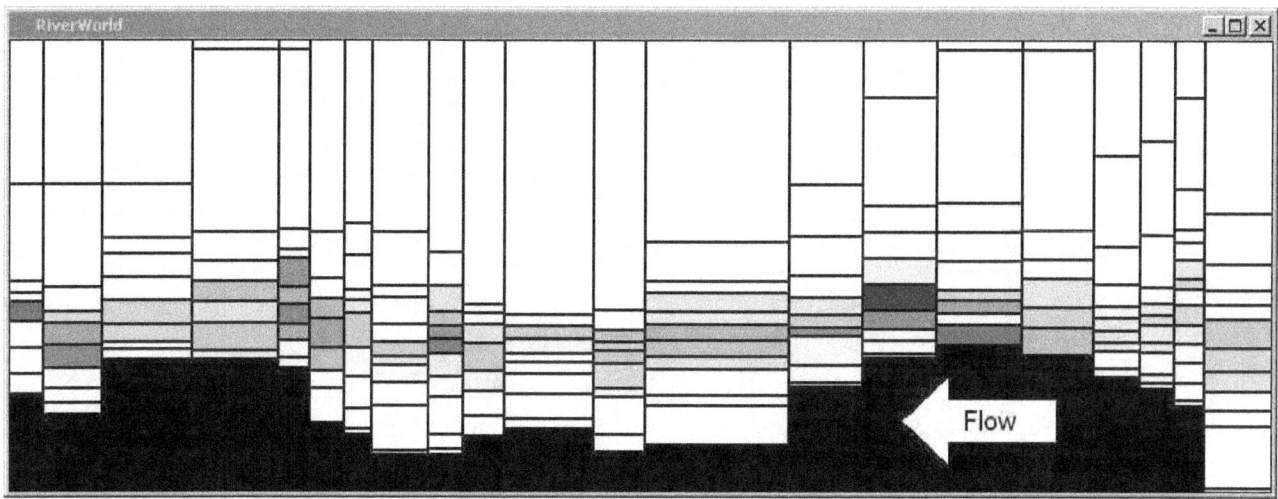

Figure 1—Representation of space in InSTREAM: nonuniform rectangular cells in rows across the river channel. This is a plan (top-down) view of one reach; flow is from right to left. White cells are dry at the depicted flow; black areas are not modeled. Cells with water are shaded by velocity, with darker grey being faster. The graphic is scaled separately in the X (upstream-downstream) and Y (across-channel) dimensions: the reach is 184 m long (X) and 47 m wide (Y).

Scheduling and temporal scales—

Time is modeled using discrete time steps. Version 4 of InSTREAM uses a 1-day (24-hour) time step, although the number of daylight hours varies with date and affects some processes. The schedule of model actions within each time step is summarized in section 2.3. Users select the temporal extent (duration) of model runs; 5- to 20-year runs are typical.

Habitat input variables—

There are three time-varying inputs to InSTREAM: daily values of water discharge, temperature, and turbidity. These variables are assumed uniform throughout a reach but can differ among reaches. There are also habitat inputs that are constant over time but variable among cells. These variables define the size and location of cells and the availability of habitat resources such as hiding and feeding cover and spawning gravel. Each cell also has hydraulic parameters that determine the cell's daily depth and water velocity from the reach's flow rate. Two variables for the availability of drift and search food (section 5.2.6) are assumed constant over both space and time, although they could be made into daily inputs via a simple software change. Habitat input also defines the location of any barriers to upstream fish movement, and how multiple reaches are linked together.

Outputs—

Unlike conventional population models, InSTREAM produces a variety of output types. One type is summary statistics of population status: abundance; mean, minimum, and maximum length; mean weight; etc., categorized by species and age class. Statistics on habitat use (e.g., histograms of fish density vs. velocity) are also produced. The model also offers output on mortality: how many fish and eggs died categorized by cause of death. Optional graphical output displays the location and size of individual fish so patterns of habitat use and movement behavior can be observed.

2.3. Trout Species and Number of Species

Through some simple software edits explained in chapter 4, InSTREAM can be made to represent any trout species and any number of coexisting species. Most model outputs are reported separately for each species, and each species has its own set of parameter values. Its object-oriented software also makes it easy to modify InSTREAM so that different species can use different methods for selected processes. The order in which individuals are processed each time step is determined only by their length, not species.

Several factors influence the number of species represented in a given application. If differences among species (e.g., in spawning season) are believed important to population dynamics, they should be represented separately. In some studies, it may be desirable to represent trout of the same species using two InSTREAM species—for example, to compare hatchery and wild trout, or to track trout from two reaches separately.

2.4. Summary of Model Actions and Schedule

The tabulation below provides a summary of the objects represented in InSTREAM and the sequence (top to bottom) of actions executed in each daily time step. Chapter 2 explains the reasoning behind this sequence of actions.

Objects	Daily Actions
Reaches	Read daily flow, temperature, turbidity.
Cells	Calculate new depth, velocity.
Female fish ready to spawn (in order from largest to smallest)	Select spawning cell, create redd, select male spawner, incur spawning weight loss.
All fish (in order from largest to smallest)	Select and move to the habitat cell that provides highest expected fitness.
	Feed and grow
	Determine survival or mortality due to predation, poor condition, and other risks
All redds	Determine egg mortality from extreme temperature, dewatering, scour, superimposition.
	Increment developmental state of eggs.
	Create new fish if eggs are fully developed.
Output graphics	Re-draw cells, fish, redds on the screen animation; update graphs.
Output files	Write daily file output.

3. Terminology and Conventions

This section describes the terms and modeling conventions followed in this document and in the InSTREAM software.

3.1. Terminology

We use the following definitions throughout this document. Much of the terminology follows Grimm and Railsback (2005).

Action. An element in an IBM's **schedule** that describes when some part of the model is executed. An action is defined by a list of model **objects** (e.g., habitat cells, fish, output files), the **methods** of these objects to be executed (e.g., **traits** of fish, updating habitat cells, producing output), and the order in which the objects are processed. For example, the fish's "move" action tells the model to use the list of all fish, to execute the "move" method for each fish, and to process the fish in order of length, from largest to smallest.

Behavior, individual behavior, system behavior. What a model fish or fish population actually does during a simulation. A behavior is an outcome of an IBM and the **traits** of its individuals.

Calibration. The process of adjusting the values of a few **parameters** to make the model reproduce field observations. Although calibration is the primary means of evaluating parameters for simple models, for InSTREAM, it is best to adopt as many parameters as possible from the literature and, subsequently, calibrate only a few (see chapter 3).

Cell. The basic unit of habitat in InSTREAM; habitat conditions differ among cells, but not within a cell. A row of cells across a stream constitutes a **transect**.

Data. **Input** that describes the habitat and fish population to be simulated. Data for InSTREAM version 4 includes daily time series of flow, temperature, and turbidity; cell dimensions and state variables; the relations between flow and depth and velocity for each cell; and the characteristics of the initial fish population.

Fish, trout. Simulated individual fish. Except where explicitly noted otherwise, these terms refer to virtual, not real, fish. Likewise, the words "egg" and "redd" refer to their virtual representation within InSTREAM.

Habitat selection. The **behavior** and corresponding **trait** for selecting which **cell** to occupy each day.

Input. Any of the **data** and **parameter** values a user provides to InSTREAM to define a **scenario**.

Method. In object-oriented software, a block of code that executes one particular **trait** or process. Methods are similar to subroutines in non-object-oriented software.

Mortality source. A natural process (e.g., starvation, predation) that causes fish or eggs to die. Mortality sources are modeled as **survival probabilities**.

Object. Something that is represented as a discrete entity with its own state variables. Example objects include individual fish, redds, and cells; and (in the software) observer tools such as graphics windows and the devices that produce output files.

Observation, observer tools. The process of collecting data and information from the IBM. Typical observations include graphical display of patterns over space and time and file output of summary statistics. Observer tools are software tools such as graphical user interfaces that make certain kinds of observation possible.

Parameter. A user-specified coefficient for one of the equations used to define traits of fish and habitat. Parameter values are one of several kinds of **input**. Compare to **variable**. Parameter values are ideally developed from empirical literature (as discussed throughout section 4) or field data. A few parameters are best evaluated via **calibration** (discussed in chapter 3).

Population. All the model fish in a simulation. (Or, for simulations with multiple species, all the model fish of a species.)

Reach. InSTREAM models the trout population in one or several reaches. Each reach is a continuous section of a stream or river channel. The habitat within a reach is broken into **transects** and **cells**.

Replicates. Multiple model runs that represent the same **scenario** but use different pseudo-random number sequences. Replicates are useful for evaluating how much of the variation in results is due to stochasticity in the initial size, mortality, and spawning effort and timing of individuals.

Scenario. A single, complete set of **input** to InSTREAM, representing one particular set of environmental conditions or one management alternative. Effects of alternative environmental conditions or management alternatives are typically assessed by comparing output produced by several different scenarios.

Schedule. A description of the order in which events are assumed to occur: the schedule defines the **actions** and the rules for executing them. In an IBM's software, the schedule is the code that defines actions and controls when they are executed.

Site. We use "site" loosely to refer to the stream or watershed area being modeled (in contrast to the specific meaning of **reach**. A site could be represented by several reaches.)

Survival probability. A model of a **mortality source**. This term refers to a fish's probability of surviving a particular kind of mortality for 1 day; but it also refers to the methods used to calculate that probability.

State, state variable. A measure of the status of some part of a model (individuals, habitat cells, the population) that typically can be described using a single number. State variables may be constant over time and be read from input **data**, or may be updated over time by model calculations. Example fish states are weight, sex, and location; cell state variables include distance to hiding cover (a constant input) and food availability (which varies daily); example system states are population biomass, number of species, and mortality rate (number of individuals dying per time step).

Submodel. A part of an IBM's **formulation** that represents one **trait** or process. Dividing InSTREAM into submodels allows each process to be modeled, calibrated, and tested separately.

Trait. A model of a particular **behavior** of individual fish. A trait is a set of rules for what individuals do at particular times or in response to specific situations in the IBM.

Transect. A straight row of **cells** across the stream and flood plain, perpendicular to the direction of flow. (In other instream flow models, the word "transect" often refers to a line across the channel along which depth and velocity are measured. This data collection approach is not necessarily appropriate for InSTREAM, so "transect" refers to a row of cells.)

Variable. Any number used in calculations. A variable may be a **parameter** or a **state variable**, or may be a temporary internal variable.

3.2. Conventions

3.2.1. Measurement units—

Distance and length are in centimeters (cm), and, therefore, areas are in square centimeters (cm^2), volumes in cubic centimeters (cm^3), and velocities in centimeters per second (cm/s). There are two important exceptions to this convention. Stream-flow is in units of cubic meters per second (m^3/s) because cm^3/s is an unfamiliar and cumbersome measure. Habitat input files that define the size and location of cells and the location of movement barriers use distances in meters (m) for convenience. However, all internal variables and outputs involving depth, velocity, area, or distance use length units of centimeters.

Weight is in grams (g).

Temperature is in degrees Centigrade (°C).

Turbidity is in nephelometric turbidity units (NTU).

Time is in days (d), because the model uses a daily time step. However, there are several exceptions to this convention. Flow and velocity variables are per second. Food availability and intake calculations use hourly rates because the number of hours per day that fish feed is variable.

Fish lengths are fork lengths in centimeters (cm).

Fish and prey (food) weight variables use wet weight.

3.2.2. Parameter and variable names—

The model's formulation uses the parameter and variable naming conventions of the Swarm software used to code the model. This convention has two benefits. First, the variable and parameter names in the formulation document can be the same as in the software. Second, the names are long and descriptive, making variables easy to identify.

Variable and parameter names typically are made by joining several words. The first word starts with a lower-case letter, and subsequent words are capitalized (e.g., *fishWeightParamA*). Input parameter names start with the kind of object that uses the parameter. These objects include fish, redds, habitat cells, fish mortality sources, and redd mortality sources. Consequently, most parameters start with the words fish, redd, cell, hab, mortFish, or mortRedd. This convention is not strictly followed for variables calculated internally by the model.

For example, the traditional length-weight relationship for fish:

$$W = aL^b$$

appears in this formulation as:

$$fishWeight = fishWeighParamA \times (fishLength)^{fishWeightParamB}$$

and the corresponding program statement in the software is:

```
fishWeight = fishWeightParamA * pow(fishLength, fishWeightParamB);
```

3.2.3. Survival probabilities and mortality sources—

A number of factors can cause fish or fish eggs to die in InSTREAM. These factors are referred to as "mortality sources." Although the word "mortality" is used in parameter names and in our text, all mortality-related calculations are based on survival probabilities. A survival probability is the (unitless) probability of surviving a particular mortality source for 1 day. (The term "mortality risk" is commonly used to mean the daily probability of dying, equal to [1 – survival probability].)

Modeling mortality as a survival probability simplifies computations and reduces the chances of error. The probability of surviving several mortality sources is calculated simply by multiplying the individual survival probabilities together. Likewise, the probability of surviving one kind of mortality for *n* days can be calculated by raising the daily survival probability to the power *n*.

3.2.4. Dates—

This model uses date input in the "MM/DD/YYYY" format (e.g., 12/07/1999). The software converts this input to the computer operating system's internal date format that automatically accounts for leap years. All input data and simulations, therefore, include leap days.

Parameters that are days of the year (e.g., spawning is allowed to occur between April 1 and May 31 of each year) are input in the "MM/DD" day format.

3.2.5. Fish ages and age classes—

InSTREAM uses the convention that fish are age 0 when born and the age of all fish is incremented each January 1. (However, if a simulation starts on January 1, the birthday is skipped.) Fish are assigned to age classes, which are used to define the initial population at the start of a model run and to report simulation results. Four age classes are used (although the number of classes can be changed via relatively simple modifications to the software; see chapter 4):

- Age 0—fish that have not yet reached their first January 1.
- Age 1—fish that have survived one January 1.
- Age 2—fish that have survived the January 1 of two years.
- Age 3+—any fish older than age 2.

3.2.6. Habitat dimensions and distances—

X and Y dimensions. The hydraulic model used for InSTREAM is one-dimensional so the model assumes the river is straight with all velocities in one direction (downstream). The X and Y values referred to here are used to define coordinates (in centimeters) of cell boundaries.

The X dimension is defined to be in the downstream-upstream direction. The origin (X = 0) is at the downstream end of a reach. The Y dimension is across the channel. To correspond with computer graphics, which place the origin (X,Y = 0,0) at the top left of the screen, Y is zero on the right bank facing downstream. On InSTREAM's graphical displays of the stream, flow is therefore from right to left.

Distances between cells. Some calculations in the model require values for the distance between two cells (e.g., for finding all the cells within a fish's maximum movement distance). The distance between two cells is calculated as the straight-line distance between the centers of the cells.

3.2.7. Logistic functions—

The survival probabilities make extensive use of logistic functions, which are useful for depicting many functions that range between 0 and 1 in a nonlinear way. The Y value of a logistic function increases from 0 to 1, or decreases from 1 to 0, as the X value increases over any range. In InSTREAM, logistic functions are defined via parameters that specify two points: the X values at which the Y value equals 0.1 and 0.9. The logistic functions are defined as:

$$S = \exp(Z) / 1 + \exp(Z)$$

where

$Z = LogistA + (LogistB \times habitatVariable)$,
$LogistA = LogistC - (LogistB \times habVarAtS01)$,
$LogistB = (LogistC - LogistD) / (habVarAtS01 - habVarAtS09)$,
$LogistC = \ln(0.1/0.9)$, and
$LogistD = \ln(0.9/0.1)$.

These equations evaluate the example survival probability *S*, given the X value *habitatVariable*. The parameters *habVarAtS01* and *habVarAtS09* are the values of the habitat variable at which survival is defined to be 0.1 and 0.9, respectively. The two X value parameters (*habVarAtS01* and *habVarAtS09* in this example) must not be equal. (Section 6.4 offers many examples of logistic functions.)

Chapter 2: Model Formulation

4. Formulation of InSTREAM: Introduction and Objectives

Chapter 2 (sections 4 through 12) describes the formulation of InSTREAM: the detailed assumptions, equations, and parameters used to implement the fundamental assumptions described above. Equally important, it also shows users **why** each of the model's detailed assumptions was chosen. The formulation describes the scientific basis for each of the submodels in InSTREAM: the literature and data that were reviewed and how they were used in the model design.

This description of the formulation generally follows an object-oriented modeling approach. First, the major kinds of objects in the model (habitat reaches and cells, fish, redds) and their traits are described. The methods used to initialize model runs are then described. Last is a description of another very important element of an IBM: the schedule that determines the order in which model events occur.

5. Habitat

Habitat is depicted in InSTREAM at three scales. The entire model can represent a network of reaches (but often just one reach is used). Reaches are habitat objects representing a continguous segment of a river or stream, and cells are objects representing the habitat units that trout occupy. A model contains one or more reaches, and each reach is made up of many cells.

Methods for collecting or otherwise assembling habitat data are discussed in chapter 3.

5.1. Reaches

5.1.1. Reach-scale variables—
The parameters used to calculate food production in each cell (section 5.2.6) are assumed uniform over a reach and constant over time. Hence, they are input as reach parameters. Two other reach-level parameters are the maximum flow at which trout will spawn (section 6.1.1) and the fraction by which velocities are reduced for trout swimming in velocity shelters (section 6.3.7).

Reaches have three variables that are updated daily from input files: flow (m^3/s), temperature (°C), and turbidity (NTU). Temperature and turbidity are assumed the same for all cells in a reach. Flow is used primarily to determine the depth and velocity in each of the reach's cells (section 5.2.2). The wetted surface area of each reach is also updated daily from the flow; the reach area is simply the sum of the areas of all cells with depth greater than zero.

Barrier locations are also reach-scale inputs. Barriers are model objects representing obstructions such as waterfalls or dams that prevent trout from moving upstream. Barriers also affect downstream movement by trout, in a more complex way (section 6.2.2). There are no limits to how many or where barriers are simulated; barriers can occur anywhere within a reach or, in simulations with multiple reaches, at either end of a reach. (Usually, barrier locations coincide with the border between two transects.) There is no capability in InSTREAM to simulate barriers that are passable by large trout but not small ones; no fish can pass upstream over a barrier, and all fish can pass downstream if they choose to. (Complete blockage of movement in both upstream and downstream directions can be simulated by building completely separate models for the reaches above and below the barrier; there is no need to combine them in one simulation.)

Barriers are assumed to cross the channel perpendicularly, so their location is defined by an X coordinate value: the distance (*m*) upstream from the downstream end of the reach. This location can range from zero, for a barrier at the downstream end of the reach, to the reach's total length, for a barrier at the upstream end of the reach.

The day length (*dayLength*, number of hours of daylight, including twilight) is a calculated reach variable. (The same day length is used for all reaches.) Day length determines the time trout spend feeding (section 6.3.2) and affects predation mortality (section 6.4). The value of *dayLength* is updated daily, using equations modified from the Qual2E water quality model (Brown and Barnwell 1987).

$$dayLength = 24 - 2\left[\left(\frac{12}{\pi}\right)\arccos\left\{\tan\left(\frac{\pi \times siteLatitude}{180}\right)\tan\delta\right\}\right]$$

where:

$$\delta = \left[\left(\frac{23.45}{180}\right)\pi\cos\left\{\left(\frac{2\pi}{365}\right)\left(173 - julianDate\right)\right\}\right]$$

and *siteLatitude* is a model parameter set to the study site's latitude (in degrees) and *julianDate* is the Julian date (day of the year [1 to 366] calculated internally from the date). This equation works only for the Northern Hemisphere.

Another reach-scale variable is the density of piscivorous trout, *piscivorousFishDensity* (number per square centimeter), used to model predation by fish (section 6.4.6). This variable is calculated as the number of piscivorous fish, divided by the reach's area. Reach area is evaluated each time step as the sum of the areas of all cells that have depth > 0 at the current daily flow. A length threshold determines if a given fish is piscivorous, and the number of piscivorous fish varies over time as fish grow and die (section 6.4.6). The value of *piscivorousFishDensity* for each reach is updated during the fish's habitat selection action: after each trout executes its habitat selection decision (in descending order of trout size), the model determines whether the trout is piscivorous and, if so, increases *piscivorousFishDensity* for the reach the trout occupies.

5.1.2. Reach links—
Users of InSTREAM specify the number of reaches and how they are linked. (Often, only one reach is used.) Reaches can be linked in a network of any kind, including a linear sequence (multiple mainstem reaches only), mainstem and tributaries, and distributaries (fig. 2).

The reach network is specified by providing, for each reach, a reach name and junction numbers for the upstream and downstream ends of the reach. The reach name is a character string up to 30 characters with no spaces. The reach name is used within the software and in output files to label each reach.

For each reach, junction numbers are provided as two reach parameters: *habUpstreamJunctionNumber* and *habDownstreamJunctionNumber*; both are integers. Junction numbers are used only to build the links that define the reach network, so their value can be arbitrary as long as they are consistent among reaches. Any two or more reaches with the same junction number will be linked at that junction. Figure 2 illustrates ways that networks of reaches can be defined, and table 1 describes how these networks are defined using junction numbers.

5.2. Cells

5.2.1. Cell boundaries and dimensions—
All the cells on one transect have the same length in the X (upstream-downstream) dimension, but differ in width, the Y (across-channel) dimension. For each transect, the user provides the X coordinate for the upstream end of the transect's cells. For each cell, the user provides the Y coordinate of the cell's right boundary. These coordinates are measured in the field. From this input, InSTREAM then calculates the boundary locations and size of each cell, which do not vary with streamflow.

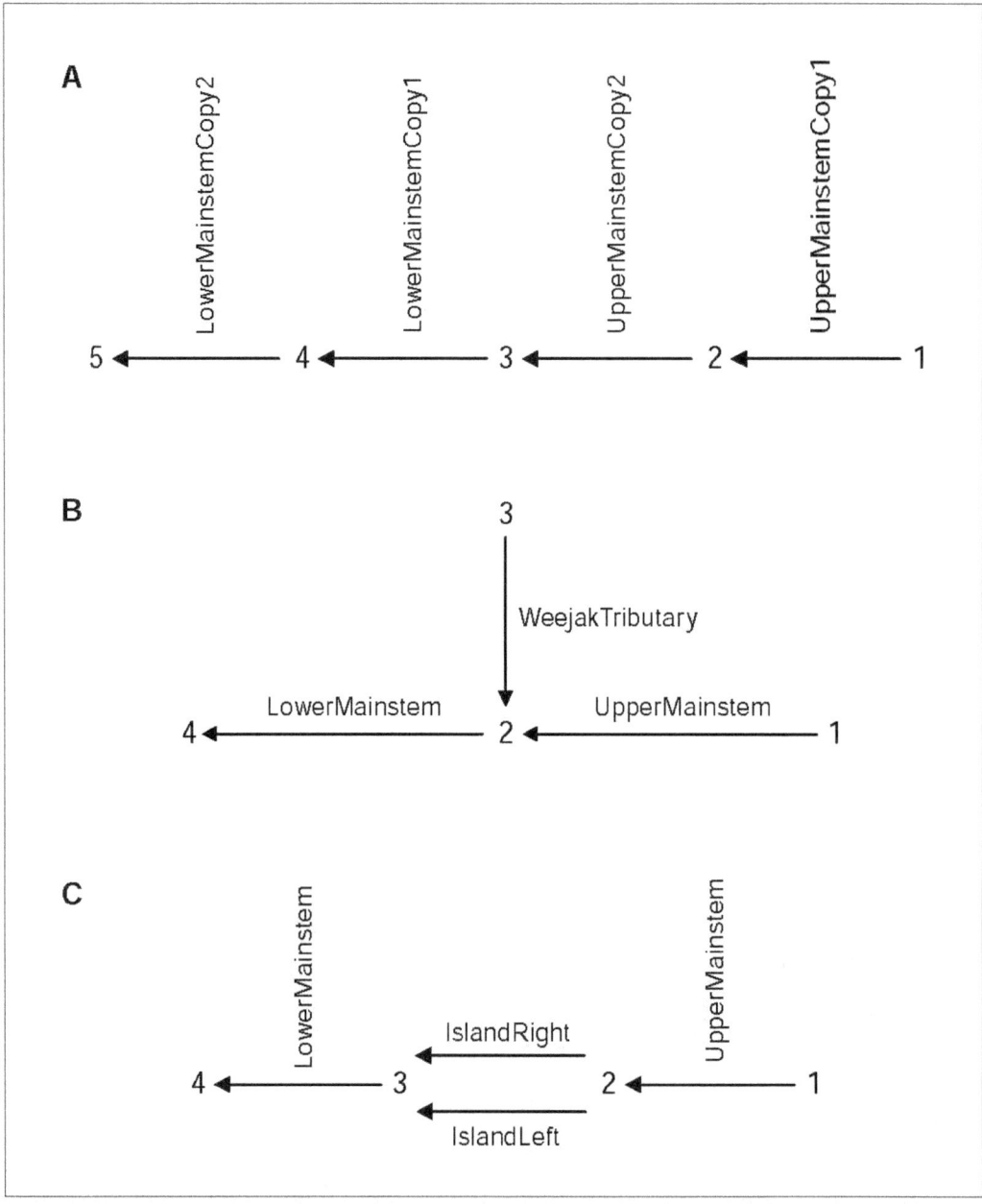

Figure 2—Example reach network configurations, showing junction numbers and reach names. Arrows represent reaches, pointing in the downstream direction. Network A has four sequential reaches generated by using two copies each of an upper and lower study site. Network B has two mainstem reaches and a tributary. Network C has reaches on either side of an island.

Table 1—Junction numbering for the example reach networks

Network	Reach name	Upstream junction number	Downstream junction number
A	UpperMainstemCopy1	1	2
	UpperMainstemCopy2	2	3
	LowerMainstemCopy1	3	4
	LowerMainstemCopy2	4	5
B	UpperMainstem	1	2
	LowerMainstem	2	4
	WeejakTributary	3	2
C	UpperMainstem	1	2
	IslandLeft	2	3
	IslandRight	2	3
	LowerMainstem	3	4

5.2.2. Depth and velocity—

The depth and velocity of each cell (and the number of cells that are submerged and therefore available to trout) vary with the daily reach flow. InSTREAM currently uses one-dimensional hydraulic models, so a cell's water velocity is treated as the speed water moves in the downstream direction.

To take advantage of existing stream hydraulic modeling software and avoid having to include hydraulic simulations, InSTREAM imports lookup tables of water surface elevation (WSE) and velocity, as a function of flow, for each cell. This approach allows all the hydraulic model building, testing, and calibration to be conducted in existing, specialized hydraulic software. The InSTREAM software imports WSE and velocity lookup tables produced by any of the hydraulic models included in the RHABSIM habitat simulation package (TRPA 1998). Methods for hydraulic simulation are discussed in chapter 3. (Other hydraulic models, including two-dimensional hydrodynamic models, could be used to produce the input needed for InSTREAM. However, new software would be needed to link model results to InSTREAM.)

The input WSE and velocity lookup tables should contain a wide range of flows and the cells' WSEs and velocities for each flow. For any flow, depth of a cell is calculated by subtracting the cell's bottom elevation from the WSE. If this differ-ence is negative, the cell is above water and depth is set to zero.

In the example depicted in figure 3, the cell is dry (depth and velocity are zero) at flows up to 20 m^3/s. As flow increases, depth increases steadily. Velocity in this example cell, however, does not increase monotonically with flow: it increases rapidly with flows between 25 and 30 m^3/s, then drops off, then increases sharply at flows around 85 m^3/s. Such discontinuities in velocity changes with flow are

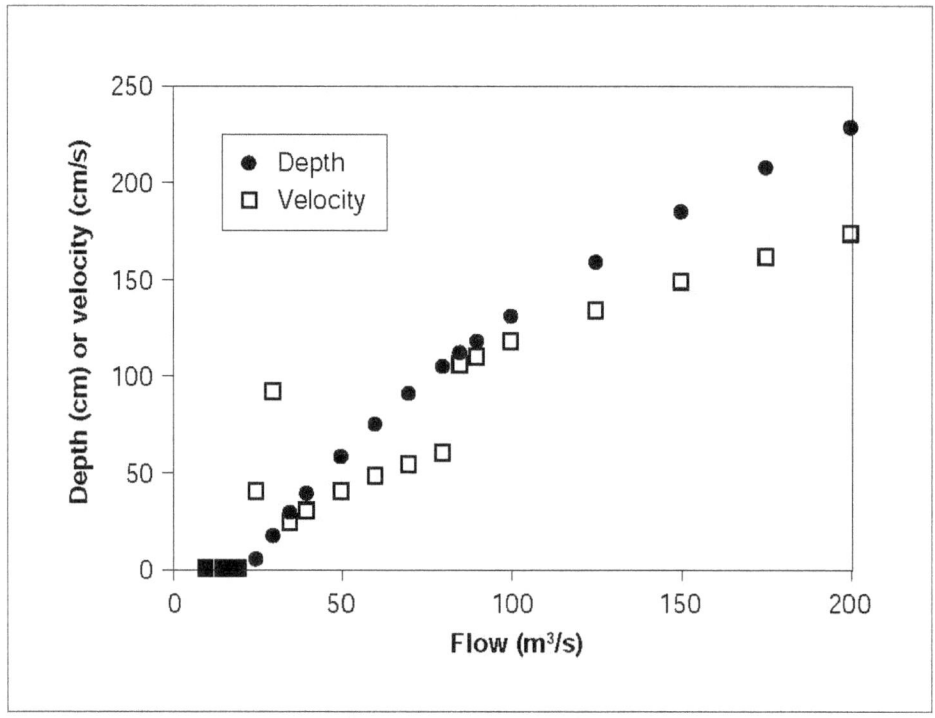

Figure 3—Example depth and velocity input for a cell. Each point represents an entry in the water surface elevation and velocity lookup table that is input for the cell.

in part an artifact of the hydraulic simulation methods (three hydraulic model calibrations were used for low, middle, and high ranges of flow), but also reflect the discontinuities that really occur in rivers. Because of eddies and other hydraulic complexities, it is not unusual for velocity to decrease in a cell as flow increases over some ranges. (This example is slightly atypical: velocity does increase monotonically with flow in most cells. Typically, velocity and depth increase nonlinearly with flow. However, exceptions like this are common; the example is presented to reinforce that capturing natural hydraulic complexity should be the highest priority in hydraulic simulation.)

On each simulation day, the WSE and velocity of each cell are computed from the reach's daily flow using the lookup tables, and depth is calculated from WSE. Linear interpolation is used, so it is important for the lookup table to include many flows. For any flows below the lowest in the lookup table, WSE and velocity are extrapolated downward linearly from the lowest two flows. Likewise, for flows above the highest in the lookup table, WSE and velocity are extrapolated linearly upward from the highest two flows in the table. The need to make these extrapolations can be avoided by making sure the lookup table includes flows lower and higher than any occurring during a model run.

At flows below the lowest in the lookup table, it is possible for this interpolation method to produce negative velocities for cells that have non-zero velocity at the lowest lookup table flow. In this case, any negative velocities are set to zero. Any channel margin cells submerged only at the highest flow in the lookup table can have unrealistically high velocities projected for flows above the highest lookup table flow. Cells submerged only at flows **above** the highest lookup table flows will have, at such flows, non-zero depth but be assigned zero velocity (because all velocities in the lookup table are zero).

5.2.3. Velocity shelter availability—

The availability of velocity shelters (which affect growth, section 6.3.7; and high velocity mortality, section 6.4.2) is modeled by assuming that a constant fraction of each cell's area provides velocity shelter. (Although the hydraulic model computes an average velocity for each cell, velocity is assumed not to be uniform within a cell.) This fraction is provided as input (variable *cellFracShelter*, a dimensionless fraction between 0 and 1) determined by field estimation. These fractions should include any part of the cell with complex hydraulics that could be used by trout to reduce their swimming speed while drift feeding. Velocity shelters can be provided by boulders, cobbles, or other substrata that induce roughness in the bottom, such as woody debris, roughness in the banks or bedrock channel, or adjacent cells with near-zero velocities. In reality, the availability of velocity shelters can vary with a fish's size and the flow, but InSTREAM ignores this variability because of its complexity. Instead, *cellFracShelter* should represent drift feeding habitat for mid-sized to large trout.

A cell keeps track of its total velocity-shelter area (*cellFracShelter* × *cellArea*) and also keeps track, over time, of how much of that shelter area is occupied by fish. Each fish using velocity shelter in a cell occupies an area of shelter equal to the square of the fish's length (section 6.3.7). A fish has access to shelter if the total shelter area of its cell is greater than the shelter area already occupied by more dominant fish. This means that a fish has access to shelter if there is **any** unused shelter space available for it in the cell. (Competition for food, not velocity shelter space, is more likely to limit the density of fish in a cell.)

5.2.4. Spawning gravel availability—

Spawning gravel availability is described as the fraction of cell area with gravel suitable for spawning, assumed to be constant over time. This area can include small pockets of gravel behind boulders as well as larger spawning beds. This spawning gravel fraction (variable *cellFracSpawn*, a dimensionless fraction between 0 and 1) is estimated in the field and provided as input for each cell.

5.2.5. Distance to hiding cover—

The habitat input variable *cellDistToHide* (m) is an estimate of how far a fish in the cell would have to move to find hiding cover. This variable is used in the terrestrial predation mortality model (section 6.4.5). The kind of habitat that trout can use for hiding differs with fish size. The terrestrial predation formulation is designed so that *cellDistToHide* should represent hiding for midsized to large trout.

5.2.6. Food production and availability—

The amount of food available to fish is a very important habitat variable, probably more important than flow or temperature in determining fish population abundance and production except under extreme conditions. Unfortunately, the processes influencing food availability for stream salmonids are complex and not well understood. Although some studies (Gowan and Fausch 2002, Morin and Dumont 1994, Railsback and Rose 1999) indicate that food availability and consumption can vary with factors including flow, temperature, fish abundance, and physical habitat characteristics, there is little information available on how food availability varies over time and space at scales relevant to individual-based models. Modeling food production is also complicated by the multiple sources of food available to fish. Stream salmonids are commonly observed feeding both by "drift feeding"— maintaining a stationary position and capturing food that drifts past; and by "search feeding"—actively searching for food on the stream bottom or surface. InSTREAM separately models "drift" food that moves with the current and "search" food that is relatively stationary and must be searched out by the fish. Both drift and search food may originate from benthic production or from terrestrial input.

Because InSTREAM assumes fish compete for the food available in each cell, cells must keep track of (a) how much food of each type is produced each day, and (b) how much is available to a particular fish.

Food production—In the absence of established models of trout food availability, InSTREAM uses models that are simple yet mechanistic and easily calibrated using observed trout growth and survival (as discussed in chapter 3). Food production is modeled using the simple assumption that (1) the concentration of food items in the drift (*habDriftConc*, grams of prey food per cubic centimeter of stream volume) and (2) the production of search food items (*habSearchProd*, grams of prey food produced per square centimeter of stream area per hour) are constant over time and space. These two variables are input as habitat parameters. [How food is produced in specific habitats such as riffles, and depleted by fish as it travels downstream, has been simulated in other models (e.g., Hughes 1992b). However, the model of Hughes (1992b) shows that simulating drift production and depletion over space would require a major increase in the complexity. The simpler approach used in InSTREAM appears to capture the important dynamics of food competition.]

The trout feeding formulation uses hourly food production and consumption rates because the number of feeding hours per day varies. The hourly food production rates are determined by the physical characteristics of habitat cells. The rate at which search food is produced in a cell (*searchHourlyCellTotal*, g/h) is simply the cell area multiplied by *habSearchProd*.

The rate at which drift food is produced in a cell (*driftHourlyCellTotal*, g/h) is modeled as the rate at which prey items flow into the cell from upstream, plus the rate at which consumed prey are regenerated within the cell:

$$driftHourlyCellTotal = 3{,}600 \times cellWidth \times cellDepth \times cellVelocity \\ \times\ habDriftConc \times (cellLength/habDriftRegenDist).$$

The constant 3,600 converts the rate from per second to per hour. The last term in this equation has two purposes. First, it simulates the regeneration of prey consumed by drift-feeding fish. Second, it makes the amount of drift food available per cell area independent of the cell's length. Without this term, five transects with cells 2 m long would have five times the food availability of one 10-m-long transect. This term keeps the amount of food available from being an artifact of how transects are spaced.

The parameter *habDriftRegenDist* (cm) should theoretically have a value approximating the distance over which drift depleted by foraging fish is regenerated. Smaller values of *habDriftRegenDist* provide higher production of food in a cell. This parameter can be used to calibrate habitat selection and starvation survival; varying it changes drift food availability without changing the amount that a drift-feeding fish captures. The parameter *habDriftConc* also affects the amount of food in a cell, but unlike *habDriftRegenDist*, also affects food capture rates of drift-feeding fish (section 6.3.3).

Estimation of values for these food parameters, including calibration, is discussed in section 6.3.10 and chapter 3.

Availability—The amount of food available to a particular trout affects the trout's habitat selection and growth (section 6.2.1). Food availability to a fish is modeled as the hourly rate at which food is produced but not consumed by larger fish. Availability is tracked separately for drift and search food; these rates are *driftHourlyCellAvail* (g/h) and *searchHourlyCellAvail* (g/h). For example, a cell's drift food may be completely consumed by larger fish (*driftHourlyCellAvail* is zero) while all of its search food remains available for any fish that chooses to use search feeding (*searchHourlyCellAvail* equals *searchHourlyCellTotal*).

The cells keep track of drift and search food availability. At the start of a simulation day, *driftHourlyCellAvail* is set equal to *driftHourlyCellTotal* and *searchHourlyCellAvail* is set equal to *searchHourlyCellTotal*. As the trout execute their habitat selection methods (section 6.2), the rate of drift or search food consumed by any fish choosing to occupy the cell is subtracted from the food availability rate for additional fish. When a fish's consumption is limited by the amount of food available in the cell, its consumption will equal all the remaining available food and no food will be available for additional fish. Any fish moving into a cell where all the (drift or search) food is consumed by larger fish will consequently have zero (drift or search) food available.

6. Fish

This section describes the methods used by the fish objects in InSTREAM. Fish are one of the two trout life stages distinctly represented in the model; the other life stage—incubating eggs and alevins—are represented by redd objects (section 7). Once fish have emerged from their redd, the methods and parameters they use do not vary with age.

Fish daily carry out four sets of actions: spawn, select a habitat cell, feed and grow, and survive or die according to survival probabilities that vary with habitat cell and fish characteristics. The methods used in these actions are described in this section. The schedule for fish actions—the order in which they are executed— is summarized in section 12.2.

Some of the parameters used in fish methods are clearly species-specific or site-specific. Example values for these parameters are provided here, along with information on the species or sites for which they were developed. Many parameter values, however, can be considered acceptable for stream trout in general; whatever variation there may be in parameter values among species is expected to be unimportant compared to other variability and uncertainty in the method the parameter is used in.

6.1. Spawning

Spawning is included in InSTREAM because the model's objectives require simulation of the full life cycle and multiple trout generations, and of the effects of flow and temperature on reproduction. Salmonids are clearly capable of adapting some of their reproductive behaviors to environmental conditions and their own state, especially by spawning or not each year depending on their current size, physiological condition, and habitat conditions (e.g., Nelson et al. 1987). However, InSTREAM's objectives do not justify a detailed representation of such processes as the bioenergetics of spawning or the adaptive "decision" of whether to spawn each year. Instead, InSTREAM's spawning methods simply force model trout to reproduce general spawning behaviors observed in real trout. Behaviors are included only if they appear important for simulating flow and temperature effects on reproduction or for representing the effects of spawning on the adult spawners.

Spawning simulations include five steps: females determine whether to spawn, select a cell to spawn in, create a redd, and identify a mate; then, both females and males incur a weight loss.

6.1.1. Determining whether to spawn—

Each day, each female trout may or may not meet the series of fish- and habitat-based spawning criteria described below. These criteria limit spawners to females of adequate size and physiological condition and restrict spawning to physical conditions (dates, flows, temperatures) when spawning by real trout has been observed, presumably because spawning is more likely to be successful during those conditions. The criteria for readiness to spawn do not include a requirement that good spawning habitat be available; it is assumed that trout will spawn whether or not ideal gravel spawning habitat is present. This assumption is supported by Magee et al. (1996).

On days when all the spawning criteria are met for a female, whether she actually spawns is determined stochastically. The probability of spawning on any such day is the parameter *fishSpawnProb* (unitless). This stochastic selection of spawning date imposes some variability in when individual fish spawn, which can be important to the population's reproductive success. Flow fluctuations during the spawning season can scour or dessicate redds of early spawners. If all spawning is early, for example, then such events can eliminate a year class of fish. The value of *fishSpawnProb* also gives the model user some control over what percentage of spawning-sized fish actually spawn. If the inverse of *fishSpawnProb* is large compared to the number of days in the spawning period (e.g., $1/fishSpawnProb$ is greater than the number of potential spawning days), then it is likely that some potential spawners will not spawn.

A value of 0.04 appears generally reasonable for *fishSpawnProb*. It causes an average of 25 percent of ready fish to spawn in the first week of suitable conditions and 68 percent to spawn within 28 days of suitable conditions.

The following subsections describe the spawning criteria. None of these criteria are well defined in the literature because trout spawning is very difficult to observe. However, the criteria make ecological sense because they keep fish from spawning at times when redds would be highly vulnerable. The criteria are included in the model for the same reason: to keep model trout from spawning under conditions that make successful incubation unlikely.

Minimum length, age, and condition—Because InSTREAM does not explicitly simulate the bioenergetics of reproduction, it uses fish length, age, and condition to predict energetic readiness to spawn. Minimum values of these characteristics are used to ensure that only fish with energy reserves comparable to those needed for gonad production can actually spawn. Length and condition are the primary indicators of spawning readiness as they are related to energy reserves, but the age minimum is useful in model runs where fish growth and condition are not well calibrated. Fish cannot spawn unless their age is at least equal to the value of the parameter *fishSpawnMinAge*, an integer age in years.

The model's fish cannot spawn until they attain a length equal to the parameter *fishSpawnMinLength* (This parameter is also a key variable in the "Expected Reproductive Maturity" fitness measure used as a basis of movement decisions; section 6.2.3.)

Finally, for a fish to spawn, its condition factor (section 6.3.1) must exceed the minimum condition factor parameter *fishSpawnMinCond* (unitless). Given that (a) the nonstandard definition of condition factor (section 6.3.1), (b) the growth formulation that makes it impossible for condition to equal 1.0 on any days when fish did not obtain at least as much energy as expended for respiration, and (c) the bioenergetics of reproduction are not explicitly represented and fish have no incentive to put on weight in anticipation of spawning, the value of *fishSpawnMinCond* is recommended to be slightly less than 1.0. We typically use a value of 0.98.

Values for *fishSpawnMinAge* and *fishSpawnMinLength* can differ considerably among sites and can often be estimated from site-specific census data. For cutthroat trout (*Oncorhynchus clarkii*) in the relatively small, infertile Little Jones Creek, Railsback and Harvey (2001) used 1 yr for *fishSpawnMinAge*. Railsback and Harvey (2001) used a value of 12 cm for *fishSpawnMinLength* on the basis of field observations and literature from similar sites. Meyer et al. (2003) provided data for cutthroat trout on how these spawning age and size parameters can vary with habitat conditions. This variation can be large. For example, Meyer et al. (2003) found trout in one large Rocky Mountain river did not mature until they were 30 cm long and 5 years old.

Not spawned this season—Trout are assumed not to spawn more than once per annual spawning season. The fish (both males and females) in InSTREAM have a boolean (yes-no) variable *spawnedThisSeason*. At the start of the first day of the spawning season, *spawnedThisSeason* is set to "no." If a fish spawns, its value of *spawnedThisSeason* is set to "yes." Females are not allowed to spawn if their value of *spawnedThisSeason* is already yes. (If a fish spawns, its value of *spawnedThisSeason* remains yes until spawning season starts again the next year.)

Date window—Salmonids generally have distinct spawning seasons. This is not surprising because time of year is an important predictor of factors that are critical to successful spawning. For example, early spring spawning may make eggs and fry more vulnerable to cold temperatures or streambed scour from high flows, but spawning too late may increase the probability of dessication, make offspring more vulnerable to high temperatures, or reduce their ability to compete with earlier-spawned juveniles. Therefore, InSTREAM fish can spawn only on days within a user-specified date window.

The date window is specified by two input parameters, *fishSpawnStartDate* and *fishSpawnEndDate*. These parameters are days in MM/DD format. (The spawning window can extend from the end of one year into the next; for example, *fishSpawnStartDate* can be 12/1 with *fishSpawnEndDate* 2/1.) Table 2 provides example values.

Table 2—Example parameter values for spawning date window

Species and site	fishSpawnStartDate	fishSpawnEndDate
Cutthroat trout, Little Jones Creek, coastal California (Railsback and Harvey 2001)	4/1	5/31
Rainbow trout, Tule River, Sierra Nevada, California (Van Winkle et al. 1996)	4/1	6/30
Brown trout, Tule River, Sierra Nevada, California (Van Winkle et al. 1996)	10/1	12/31

Temperature range—Temperature is widely accepted as a factor controlling the timing of spawning (e.g., Lam 1988). Temperature could be used by spawners as a cue for seasonal changes and to avoid temperature-induced egg mortality. Therefore, spawning in InSTREAM can only occur within a range defined by parameters for maximum and minimum spawning temperatures for spawning. Table 3 provides parameter values developed by Van Winkle et al. (1996).

Table 3—Parameters and example values for spawning temperature range

Parameter	Definition	Rainbow trout value	Brown trout value
fishSpawnMinTemp	Minimum temperature at which spawning occurs (°C)	8	4
fishSpawnMaxTemp	Maximum temperature at which spawning occurs (°C)	13	10

Source: Van Winkle et al. (1996)

Flow limit—High flow also limits spawning. During unusually high flow, cells with depths and velocities suitable for redds (section 6.1.2) are likely to be along river margins where redds can be dewatered when flows recede. The high flow limit is defined by a single habitat reach parameter, *habMaxSpawnFlow* (m^3/s). This is a highly site-specfic habitat parameter instead of a fish parameter because it differs among reaches.

Steady flows—Fish are assumed not to spawn when flows are unsteady because flow fluctuations place redds at risk of dewatering or scouring mortality. The parameter *fishSpawnMaxFlowChange* (unitless) is used to define this criterion: if the fractional change in flow from the previous day is greater than the value of *fishSpawnMaxFlowChange* then spawning is not allowed. This fractional change in flow is evaluated as:

$$fracFlowChange = abs(reachFlow - yesterdaysFlow) / todaysFlow$$

where *reachFlow* is the current day's flow, *yesterdaysFlow* is the flow on the previous day and abs( ) is the absolute value function. Van Winkle et al. (1996) and Railsback and Harvey (2001) estimated 0.20 as a reasonable value for *fishSpawnMaxFlowChange*.

6.1.2. Select spawning cell and move there—

Female spawners select the cell in which they then build a redd. Selection of habitat for foraging is modeled very mechanistically (section 6.2), but selection of spawning habitat is modeled in a simple, empirical way, with spawning cells chosen using preferences observed in real trout for depth, velocity, and substrate. This decision was made because a detailed, mechanistic representation of spawning habitat selection would require considerable additional complexity: modeling processes such as intergravel flow and water quality, which are extremely data-intensive and uncertain. This additional complexity is not necessary to meet InSTREAM's objectives (section 6.1), but we do need a simple representation of how flow affects where redds are placed because a redd's location affects its survival of dewatering (section 7.1.1).

The first step in identifying the location of a new redd is identifying potential spawning sites. This step uses exactly the same method used by trout to identify potential destination cells during habitat selection (section 6.2.2), with one exception: spawners will not cross barriers in either direction. If there is a barrier between a spawner's current cell and another cell, that cell is excluded as a potential spawning cell, whether the barrier is upstream or downstream of the spawner's current cell. The reason for excluding cells downstream of a barrier is to avoid rules for when a spawner would go over a barrier. Because InSTREAM does not model the effects of habitat on redd success in much detail, this simplification is not expected to have important effects.

For simulations with multiple habitat reaches, the potential spawning sites could include cells in a different reach from the spawner's current cell. Cells in another reach could be chosen for a redd even if the habitat criteria for spawning (section 6.1.1) are not all met in that other reach. For example, a female can "decide" to spawn only when habitat criteria such as temperature (section 0) are met in its current reach, but the female could then spawn in a reach where the temperature criterion is not met. This possibility remains in InSTREAM only because it was judged not important enough to justify the additional logic and computation to prevent it.

This formulation does not cause, or allow, long spawning migrations. In most applications of InSTREAM, the reaches are expected to be too few and small to represent long-distance migrations.

After potential spawning cells are identified, they are rated by the spawner to identify the cell where the redd will be created. The spawning cell is the potential spawning cell with the highest value of variable *spawnQuality* where:

$$spawnQuality = spawnDepthSuit \times spawnVelocitySuit \times spawnGravelArea.$$

The variables *spawnDepthSuit* and *spawnVelocitySuit* are unitless habitat suitability factors determined using methods described below. The value of *spawnGravelArea* is the cell area times its fraction with spawning gravel (*cellArea* × *cellFracSpawn*). (The units of *spawnQuality* are therefore cm^2, but they are not indicative of just spawning area.) The variable *spawnGravelArea* is included in *spawnQuality* because a spawner is assumed more likely to spawn in a cell that has more area of gravel, even if it does not select for bigger patches of gravel. Redd mortality from superimposition (section 7.1.5) is likely to result from this formulation because spawners search many cells for the best spawning habitat—so it is likely that more than one spawner will use the same cell. However, the best cell for spawning can vary from day to day as flow varies.

It is possible that none of the potential spawning cells have a value of *spawnQuality* greater than zero, especially where spawning gravel is extremely sparse. If *spawnQuality* is zero for all potential spawning cells, then the model assumes a spawner will still spawn but ignore gravel area as a criterion. In this situation, the spawner selects the cell with the highest value of *spawnQuality* ignoring spawning gravel:

$$spawnQuality = spawnDepthSuit \times spawnVelocitySuit.$$

If there are still no cells with *spawnQuality* greater than zero, then the spawner places its redd in its current cell. (This condition should occur very rarely, especially if *habMaxSpawnFlow* is well-chosen.)

When the female spawner has selected its spawning cell, the spawner moves to that cell. (The only effect this has on the spawner is that when it executes its habitat selection action later the same day, it will start from the cell it spawned in.) Male spawners are not assumed to move to the spawning cell.

The suitability factors *spawnDepthSuit* and *spawnVelocitySuit* are unitless variables representing the tendency of salmonids to select fairly well-defined ranges of depth and velocity for spawning (e.g., Knapp and Preisler 1999). Presumably, real trout select these ranges because they correspond to hydraulic conditions under which egg survival is generally high. For example, intermediate depths have highest suitability, likely because redds placed in shallow water are susceptible to dewatering if flows decline, and redds in deep water are more vulnerable to scouring during high flows or siltation during low flows. Intermediate velocities have highest suitability, presumably because low velocities provide inadequate flow of water through the redd (important for providing oxygen and removing wastes) and high velocities present a risk of scouring. Depth and velocity suitability functions are certainly a simplification of how salmonids select spawning habitat, but they are an appropriate simplification for InSTREAM and available in the literature for a variety of species and sites (e.g., Gard 1997).

The spawning suitability factors for depth and velocity are interpolated linearly from suitability relations provided as parameters. Values of *spawnDepthSuit* are interpolated from the parameters in table 4 (also plotted in fig. 4), which are example values for relatively small stream trout. These parameter values were estimated from a collection of rainbow trout (*Oncorhynchus mykiss*) and brown trout (*Salmo trutta*) spawning criteria (PG&E 1994). The number of points in this suitability relationship is fixed at five.

Table 4—Example parameter values for spawning depth suitability

Parameter name	Parameter value (depth)	Parameter name	Parameter value (unitless suitability)
	Centimeters		
fishSpawnDSuitD1	0	fishSpawnDSuitS1	0
fishSpawnDSuitD2	5	fishSpawnDSuitS2	0
fishSpawnDSuitD3	50	fishSpawnDSuitS3	1.0
fishSpawnDSuitD4	100	fishSpawnDSuitS4	1.0
fishSpawnDSuitD5	1000	fishSpawnDSuitS5	0

Note: The value of *fishSpawnDSuitD1* is a depth; the value of *fishSpawnDSuitS1* is the corresponding suitability value; *fishSpawnDSuitS2* is the suitability for the depth specified by *fishSpawnDSuitD2*, etc.

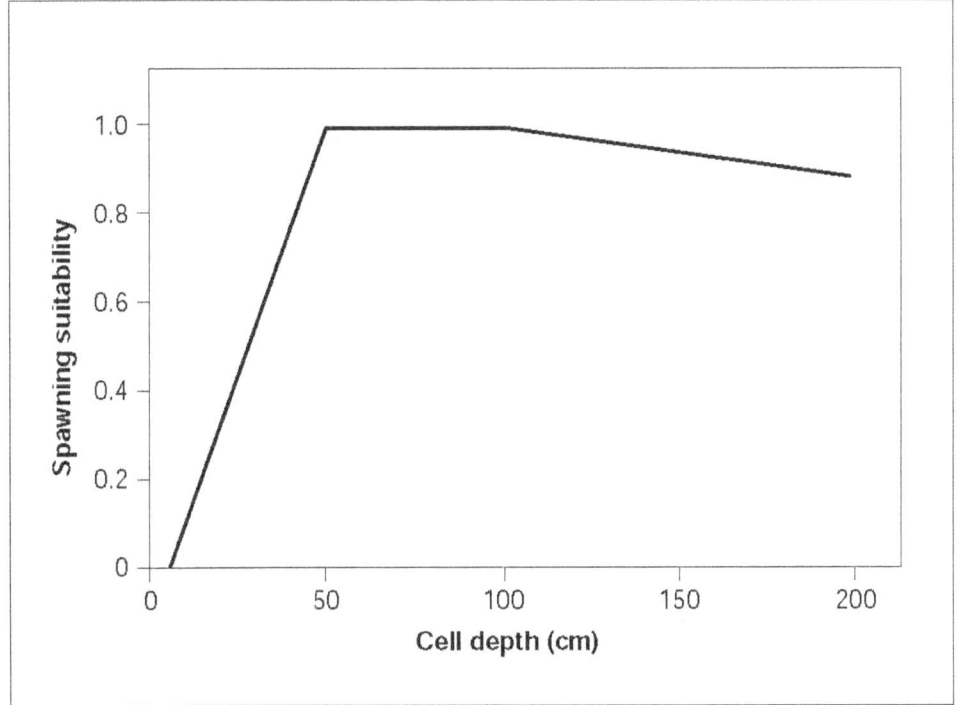

Figure 4—Spawning suitability function for depth.

A value of *spawnVelocitySuit* for a cell is interpolated from the five pairs of parameters in table 5, which includes example parameter values for small trout. The parameter values in table 5 (plotted in fig. 5) were estimated from several brown trout spawning criteria (PG&E 1994). The number of points in this relationship is fixed at six.

Table 5—Example parameter values for spawning velocity suitability

Parameter name	Parameter value (velocity)	Parameter name	Parameter value (unitless suitability)
	Centimeters per second		
fishSpawnVSuitV1	0	fishSpawnVSuitS1	0
fishSpawnVSuitV2	10	fishSpawnVSuitS2	0
fishSpawnVSuitV3	20	fishSpawnVSuitS3	1.0
fishSpawnVSuitV4	75	fishSpawnVSuitS4	1.0
fishSpawnVSuitV5	100	fishSpawnVSuitS5	0
fishSpawnVSuitV6	1000	fishSpawnVSuitS6	0

Note: The value of *fishSpawnVSuitS1* is the suitability corresponding to the velocity specified by *fishSpawnVSuitV1*, etc.

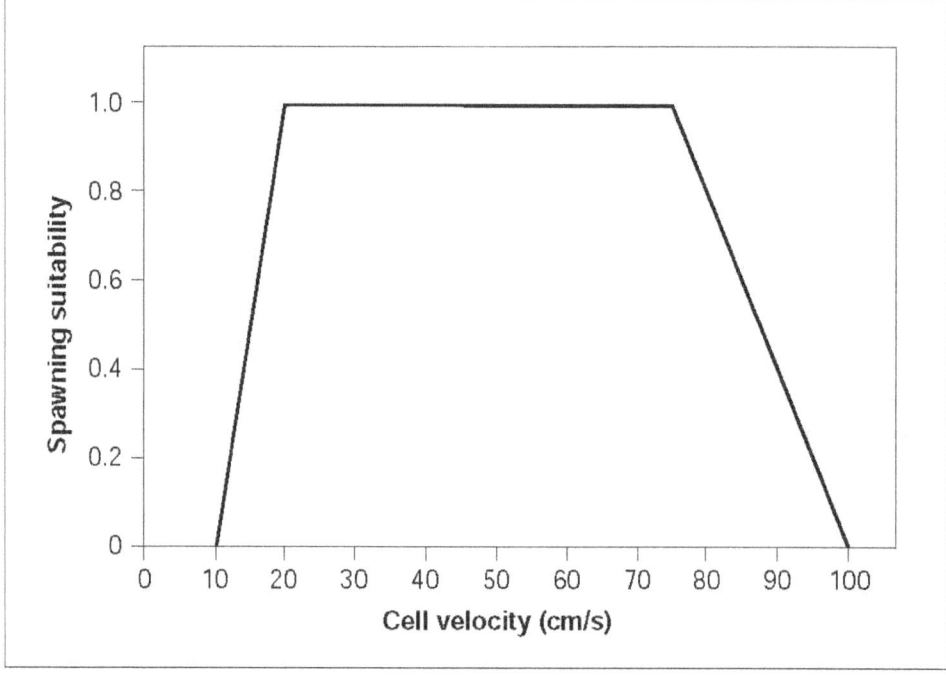

Figure 5—Spawning suitability function for velocity.

These example parameter values should be reconsidered for each application of InSTREAM. In bigger rivers, for example, greater depths may be suitable without risk of scouring; larger spawners and greater spawning gravel size may reduce the risk of scouring, making higher velocities suitable. To keep spawners from selecting cells with large areas of marginal spawning suitability instead of small cells of high suitability, it is desirable for the suitability relations to be steep-sided instead of having wide ranges of intermediate suitability.

If the model needs to interpolate a value of *spawnDepthSuit* for a depth greater than the value of *fishSpawnDSuitD5* (or a value of *spawnVelocitySuit* for a velocity greater than *fishSpawnVSuitV6*), the value is extrapolated from the last two points in the suitability relation. However, suitability values less than zero are converted to zero. Suitability values greater than 1.0 are allowed, so suitability could be scaled from 0 to 10 instead of 0 to 1.0. (It is actually very unlikely that depth and velocity have exactly equal effects on redd location, so they should have different maximum suitability values.)

6.1.3. Create a redd; set number of eggs—

Female spawners create redds after selecteding a spawning cell. The number of eggs in the redd depends on the spawner's fecundity (a function of length) and losses during spawning:

$$numberOfEggs = (fishFecundParamA \times fishLength^{fishFecundParamB}) \times fishSpawnEggViability$$

The first term in this equation is the spawner's fecundity, the number of eggs it produces. Van Winkle et al. (1996) developed values of *fishFecundParamA* and *fishFecundParamB* for brown trout from Avery (1985), which appear generally useful for relatively small stream-resident trout. These values (table 6, brown trout values) result in fecundities of 60 eggs for a small spawner of 12 cm and 220 eggs for a spawner of 20 cm, corresponding well with citations provided by Carlander (1969). Meyer et al. (2003) developed parameters for fecundity from 26 observations of resident cutthroat trout, with lengths between 10 and 30 cm. The total lengths reported by Meyer et al. were converted to fork length by applying a ratio of 0.97 (Carlander 1969). The resulting parameter values (table 6, cutthroat trout values) produce fecundities approximately 50 percent higher than those of Van Winkle et al. (1996). The differences between the two parameter sets reported in table 6 may be more a result of random variation or differences among sites than real differences among trout species.

Table 6—Parameter values for fecundity

Parameter	Definition	Species	Value
fishFecundParamA	Fecundity multiplier (eggs per redd)	Brown[a] Cutthroat[b]	0.11 0.18
fishFecundParamB	Fecundity exponent (unitless)	Brown[a] Cutthroat[b]	2.54 2.51
fishSpawnEggViability	Fraction of female's eggs that become viable eggs in the redd (unlikely to vary with species)		0.8

[a] Source: Van Winkle et al. (1996)
[b] Source: Meyer et al. (2003)

The second term consists of the parameter *fishSpawnEggViability*, which is the fraction of eggs that are successfully fertilized and placed in the redd. (Even though *fishSpawnEggViability* has the same effect mathematically as *fishFecundParamA*, fecundity and egg viability are treated separately to allow clear use of the extensive literature on fecundity.) The number of viable eggs in a redd can be considerably less than the female's fecundity if some eggs are washed away, incompletely buried, or eaten by other fish during redd creation, or if some are not fertilized. This parameter can also be used to represent mortality of eggs and alevins not explicitly included in the model (section 7.1). There is little published literature to support consistent values of *fishSpawnEggViability* for stream salmonids. For example, Healey (1991) reviewed egg deposition for Chinook salmon (*Oncorhynchus tshawytscha*) and found only a few conflicting studies, concluding that egg loss could be high in high-velocity streams but is often low. Anecdotal evidence from salmon and trout in coastal California suggests the number of emerging fry often ranges down to 50 percent of the female's fecundity.

6.1.4. Select a male spawner—

When a female spawns, InStream attempts to select a male that also spawns. The only purpose of identifying a male spawner is to impose spawning weight loss (described below) on the male. The selected male spawner is the largest fish in the simulation that meets all the male spawner criteria listed below. The largest eligible male is chosen because larger males are assumed more likely to be sexually mature (Meyer et al. 2003) and more likely to compete successfully to fertilize females (e.g., Jones and Hutchings 2002).

This selection of a male occurs after the female creates the redd. If several females spawn on the same day, the male selected for the first female spawner becomes ineligible for the subsequent female spawners on the same day. If no male meets the spawning criteria, there is no effect on the female or redd. The female still produces a fertile redd and incurs weight loss due to spawning. This assumption is made because spawning failure owing to absence of males is considered too rare and unpredictable to include in the model. Males are not assumed to move as a result of spawning.

Each selected male spawner is the largest male that:

- Is of the same species as the female;
- Occupies the same reach as the female's new redd;
- Has length greater than the parameter *fishSpawnMinLength*;
- Has age equal to or greater than the parameter *fishSpawnMinAge*;
- Has condition greater than the parameter *fishSpawnMinCond*; and
- Has not previously spawned during the current spawning season.

6.1.5. Incur weight loss; spawning mortality—

Trout lose body mass and total energy content during spawning. Lien (1978; see also Hayes et al. 2000) measured mean mass losses from spawning for brown trout of 22 percent in females and 15 percent in males, which corresponded to losses in energy content of 48 and 44 percent, respectively. Such losses can significantly affect the habitat selection and survival (especially of starvation) of spawners, so they are represented in InSTREAM. When any trout—male or female—spawns, their weight is reduced according to the parameter *fishSpawnWtLossFraction*. Fish weight is multiplied by (1 − *fishSpawnWtLossFraction*). Results from Lien (1978) suggest a value of 0.2 for *fishSpawnWtLossFraction*. However, a higher value could be justified because, as noted above, proportional energy loss substantially exceeds mass loss.

Previous versions of InSTREAM (Railsback and Harvey 2001) and previous trout IBMs (Van Winkle et al. 1996) represented the energetic effect of spawning (plus additional risks of predation and disease associated with spawning) as a single mortality risk: fish that spawned incurred a one-time risk of mortality instead of a weight loss. Spawning can result in low survival in some salmonid populations. The review by Stearley (1992) concluded that survival of spawning ranges from zero in anadromous Pacific salmon to low (5 to 50 percent) in anadromous steelhead (*O. mykiss*) and cutthroat trout, to high (80 to 90 percent) in stream trout. Vinyard and Winzeler (2000) found 10 to 35 percent survival in a population of large cutthroat trout that migrate to spawn in a small stream. These figures are for survival of the entire spawning migration, a period of several to many weeks, so they likely reflect increased swimming cost, agonistic interactions during mating, and exposure to predation in addition to consequences of energy loss. Not all of these processes are important for the resident trout modeled by InSTREAM, so the effect of spawning on survival is represented only by the weight loss.

Figure 6 represents how the probability of surviving poor condition (starvation and disease; section 6.4.4) for 90 days varies with the value of *fishSpawnWtLoss-Fraction*, using the feeding, growth, and survival parameters described in this report for cutthroat trout. The figure indicates that a 20-percent loss of body weight during spawning reduces the probability of surviving for 90 days by about 10 to 15 percent, and a 30 percent weight loss reduces survival by about 40 percent. These values are similar to the spawning survival ranges suggested by Stearley (1992) for resident trout.

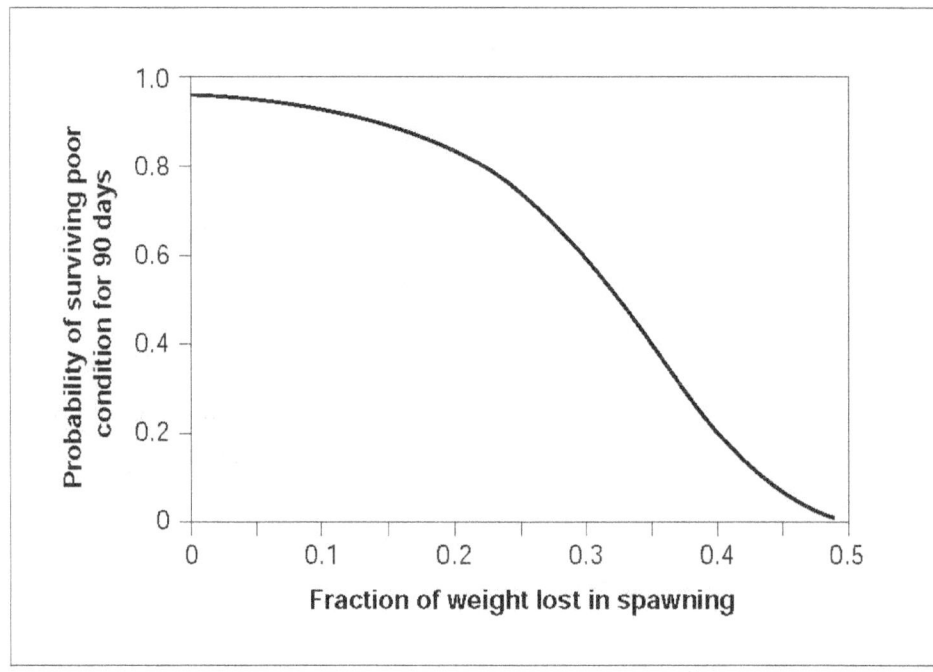

Figure 6—Probability of surviving poor condition for 90 days, as a function of spawning weight loss. A 15-cm trout feeding in velocity of 40 cm/s with velocity shelter is represented.

6.2. Habitat Selection

Habitat selection must be simulated realistically because it is probably the most important way that stream fish can adapt to short- and mid-term changes in habitat and physiological state. Therefore, modeling habitat selection has been a primary research focus in the development of InSTREAM. (The word "movement" is commonly also used for this trait; "habitat selection" is a more precise term but in this document the terms are generally interchangeable.) Railsback et al. (1999) reviewed methods used in previous models and developed the new approach used in InSTREAM. The following principles provide the basis for modeling habitat selection.

- The model will be most general and powerful if realistic habitat selection emerges when model fish use simple decision rules for responding to realistic information about environmental variation. Restrictions that force the model to reproduce a specific observed behavior that is not robust (for example, habitat "preferences" or fixed territory sizes) should be avoided.
- Stream fish are generally aware of their surrounding environment and are able to make habitat selection decisions within the model's 1-day time step.
- Habitat selection decisions must incorporate both food intake and mortality risks.

- Habitat selection can be assumed to maximize some direct measure of an animal's fitness.
- Because fitness includes the future, fitness-based habitat selection methods must consider outcomes predicted over some future period. It is assumed that an animal makes its decisions considering more than the immediate, same-time-step outcome.
- Habitat selection is strongly affected by competition among trout. Competition is best represented as a size-based dominance hierarchy.
- Whereas many other models have assumed trout compete for territory space (e.g., Gowan and Fausch 2002, Van Winkle et al. 1998), using food instead has key advantages. Territoriality appears to be a mechanism of competition for food, and territory size (and in fact whether trout defend territories instead of aggregating at particularly productive sites) can depend both on how much food is available (e.g., Dill et al. 1981) and feeding conditions (velocity, light level, turbidity) (e.g., Valdimarsson and Metcalfe 2001). Modeling competition for food instead of space avoids the need to represent how territory size varies with such factors.

The habitat selection trait resulting from these principles is conceptually simple: every day, each trout moves to the habitat cell that (1) is close enough that the fish can be assumed to be aware of conditions in it and (2) offers the highest "expected fitness," where expected fitness is approximated as the expected probability of surviving and reaching reproductive size over a future time horizon.

The habitat selection trait used in InSTREAM has been explored and tested thoroughly and found to have many unique capabilities (Railsback and Harvey 2002, Railsback et al. 1999). For example, the following observed behaviors are reproduced in InSTREAM, even though none of them are explicitly included in the model's software:

- Under "normal" conditions with food availability generally uniform over space, food competition causes trout to space themselves in a way resembling territoriality, with the area occupied per trout resembling observed territory sizes (Grant and Kramer 1990) and trout density increasing with food availability (as observed by Dill et al. 1981).
- When food availability is highly nonuniform, trout are much denser in the "hot spots" of high growth potential.
- A cell that can support only one or two large adults often also contains a few much smaller trout.
- Juveniles take greater risks than adults to grow.

- When events that alter the spatial distribution of mortality risk and growth (e.g., a flood) occur, trout respond by moving to relatively good habitat under the new conditions.
- Increased food competition causes less-dominant fish to shift habitat. Consequently, environmental changes that favor large trout often have negative secondary effects on small trout.
- The presence of piscivourous fish causes small trout to shift into shallower habitat.
- Trout generally use higher velocities in summer than in winter.
- When food availability is reduced, trout shift to habitat that has lower survival probability but higher food intake.

The following subsections explain the habitat selection trait in detail.

6.2.1. Competition for resources via dominance hierarchy—

The habitat selection trait assumes a size-based dominance hierarchy: fish can only use resources (food and velocity shelters) that have not been used by larger fish. Hughes (1992a) showed that stream salmonids rank feeding positions by desirability, and the most dominant fish obtain the most desirable sites. Gowan and Fausch (2002) and Hughes (1992a) also showed that dominance is usually proportional to length. The hierarchy is implemented in InSTREAM by executing the habitat selection method in order of descending fish length. The longest individual selects its cell first, and the food and velocity shelter it uses are subtracted from those available in the cell for additional trout. Subsequent trout therefore base their habitat selection not on the total resources in each cell but on the resources remaining unconsumed by larger fish.

Two elements of competition for food or space are not included in InSTREAM. Some literature indicates that there may be inherent interspecific differences in dominance: individuals of one species may outcompete larger individuals of another species (e.g., Magoulick and Wilzbach 1999, Volpe et al. 2000). In some earlier trout IBMs (Van Winkle et al. 1996; earlier versions of InSTREAM), the relative dominance of an individual could be a function of its species as well as its length. Similarly, some literature indicates that individuals have an inherent tendency to stay in one location ("site fidelity") and that prior residence of a site increases the ability of a trout to defend the site from larger competitors (Cutts et al. 1999, Johnsson et al. 1999, Volpe et al. 2000). However, neither species nor prior residence effects on dominance are clearly universal, and it is possible for them to be reproduced in an IBM without being explicitly programmed. For example, one species may appear to out-compete another simply because it spawns earlier in the year and so has a size advantage. Young (2003), for example, found size to be the dominant factor determining dominance among a mix of coho salmon (*Oncorhynchus kisutch*) and steelhead. Large trout may appear to exhibit site fidelity simply because their habitat offers very high fitness under a wide range of flows and temperatures, so they rarely have incentive to move. These two elements of competition are not explicitly included in InSTREAM because they are not clearly important and because doing so would require poorly founded assumptions and parameters.

6.2.2. Identify potential destination cells—

When each individual trout begins to select habitat, its first step is to identify potential movement destinations. Distance, barriers, and depth can limit potential destination cells, but the number of fish already in a cell does not limit its availability as a destination.

Distance limitation—Only habitat cells within a certain distance are included as potential destinations. This maximum movement distance should be considered the distance over which a fish is likely to be aware when desirable destinations are available over a daily time step. The maximum movement distance should **not** be considered the maximum distance a fish could swim or migrate in a day.

The maximum movement distance is a function of fish length. Because mobility and spatial knowledge are assumed to increase rapidly with fish size, we use an exponential function. The parameters *fishMoveDistParamA* and *fishMoveDistParamB* are potentially site-specific: fish are likely to explore and be familiar with larger areas in lower-gradient rivers (Diana et al. 2004).

$$maxMoveDistance = fishMoveDistParamA \times fishLength^{fishMoveDistParamB}$$

In InSTREAM, fish can follow a gradient in habitat quality if the gradient is detectable within the *maxMoveDistance*, but they do not have the ability to find and move toward some specific target if that target is beyond *maxMoveDistance*. For example, if habitat generally improves in an upstream direction, fish will have an incentive to gradually move upstream. However, if a very good location for some fish exists farther away than its *maxMoveDistance*, the fish will not be aware of it.

Movement observations from the literature cannot be considered direct measurements of *maxMoveDistance* but can be useful for evaluating its parameters. Observed movement distances (Bowen 1996, Gowan and Fausch 1996, Harvey et al. 1999) show how far fish actually move, not the distance over which they evaluate habitat. These observations are also potentially confounded by a number of factors. Small fish may actually move more than large fish because they are less able to defend a location; this does not mean small fish have a larger maximum movement distance as defined in the model. Movement rates (m/d) reported in the literature are also potentially deceptive because they are rarely based on continuous or even daily observations of location.

However, literature observations indicate that adult trout commonly move distances up to 300 m. Harvey et al. (1999) showed fall and winter movements of adult (18 to 24 cm length) cutthroat trout of up to 55 m in one day in a stream with a 1.8 percent gradient. Summer conditions (lower flows, higher metabolic rates and food requirements, higher population densities) may encourage greater movement distances. June (1981) observed little movement in newly emerged cutthroat trout <3 cm; dispersal started after they exceeded 3 cm in length. Diana et al. (2004) observed large brown trout that routinely moved between stream locations more than 500 m apart.

Parameter values for a midsized, moderate-gradient stream (table 7; fig. 7) estimate *maxMoveDistance* as less than 2 m for newly emerged trout with length of 3 cm, 5 m for juveniles 5 cm long, 30 m for trout 10 cm long, and 80 m for trout 20 cm long.

Table 7—Example parameter values for fish movement distance

Parameter	Definition	Value
fishMoveDistParam A	Multiplier for maximum movement distance (unitless)	20
fishMoveDistParam B	Exponent for maximum movement distance (unitless)	2

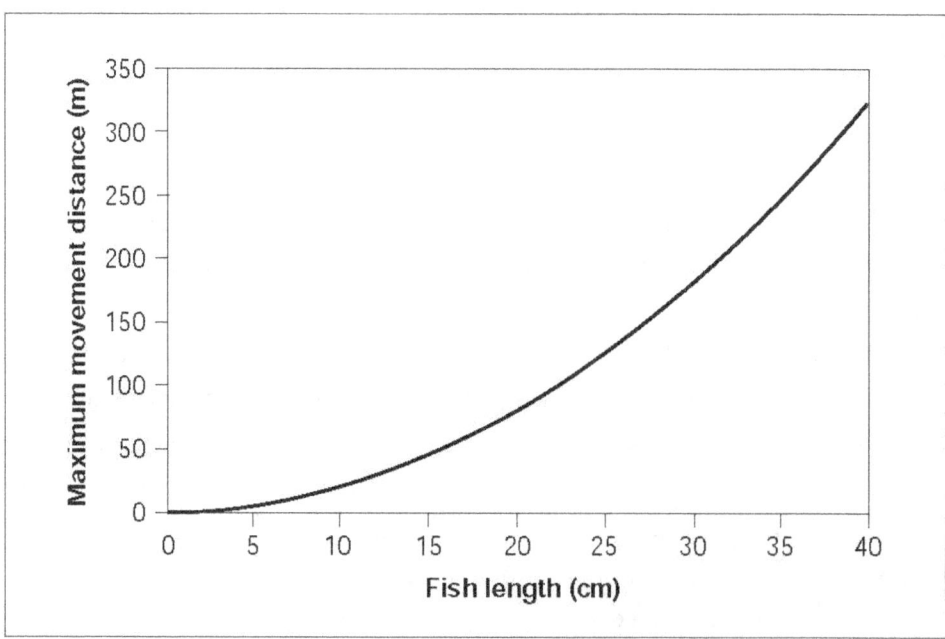

Figure 7—The maximum distance fish can move, as a function of their length, for *fishMoveDistParamA* = 20, *fishMoveDistParamB* = 2. Note the Y axis is in meters, not cm.

To identify the cells that meet the distance criterion for potential destinations, the model first calculates a trout's current *maxMoveDistance*. Second, each other cell in the trout's reach is evaluated: if the distance between a cell and the trout's current cell is less than *maxMoveDistance*, then the cell meets the distance criterion.

If a simulation includes more than one habitat reach, then cells in adjacent reaches may also be potential movement destinations for a trout. If, for example, *maxMoveDistance* for a fish is greater than the distance from the fish's current cell and the downstream end of its reach, and another reach is linked to the downstream end of the fish's reach, then some cells in the linked reach will be potential movement destinations.

The following steps are used to create a list of all cells that meet the distance criterion (*destCellList*), falling within *maxMoveDistance* of a fish's current cell (*myCell*). When determining which cells from adjacent reaches should be included in *destCellList*, only the X coordinate (which increases from downstream to upstream) is considered. These methods do not allow the possibility of a fish moving from its current reach (*myReach*) through a second reach and into a third.

1. All the cells of *myReach* within the distance *maxMoveDistance* are added to *destCellList*.

2. A test is conducted to determine whether *maxMoveDistance* extends beyond *myReach*'s downstream end. This test is true if *maxMoveDistance* is greater than the X coordinate of *myCell*'s midpoint. If this test is true, then steps 3 and 4 are conducted.

3. If the downstream end of another reach links to the downstream end of *myReach* (e.g. the end of a tributary), then cells within distance *D* of the downstream end of the linked reaches are added to *destCellList*. *D* is equal to *maxMoveDistance* minus the X coordinate of *myCell*'s midpoint.

4. If there are reaches whose upstream ends are linked to *myReach*'s downstream end, then cells within distance *D* of the upstream end of the linked reaches are added to *destCellList*. *D* is equal to *maxMoveDistance* minus the X coordinate of myCell's midpoint.

5. A test is conducted to determine whether the distance *maxMoveDistance* extends beyond *myReach*'s upstream end. This test is true if *maxMove-Distance* plus the X coordinate of *myCell*'s midpoint is greater than *myReach*'s total length in the X dimension. If this test is true, then steps 6 and 7 are conducted.

6. If there are reaches whose downstream ends are linked to *myReach*'s upstream end, then cells within a distance D of the linked reaches' downstream end are added to *destCellList*. In this case, D is equal to *maxMoveDistance* plus the X coordinate of *myCell*'s midpoint, minus the total length of *myReach* in the X dimension.

7. If there are reaches whose upstream ends are linked to *myReach*'s upstream end, then cells within a distance D of the upstream end of the linked reaches are added to *destCellList*. D is now equal to *maxMoveDistance* plus the X coordinate of *myCell*'s midpoint, minus *myReach*'s total length in the X dimension.

Steps 3, 4, 6, and 7 use methods that identify all the cells that are within a distance D of the upstream or downstream end of their reach. The method for doing so considers only the X dimension: cells are included if the distance from their midpoint to the reach end, measured in the X direction, is less than D.

For small fish, it is possible that no cells (other than its current one) are closer than *maxMoveDistance*. Having no potential destination cells poses an artificial barrier to movement, an artifact of the model's spatial resolution. This artifact could be important, for example by preventing newly emerged fish from moving from their natal redd to habitat where survival probabilities are higher. In such a situation, competition among newly emerged fish for food would largely be an artifact of the cell's size, which controls how much food is in it. To address this problem, a fish's potential destinations always include the four cells adjacent to the fish's current cell. (These four cells are identified as the cells just upstream, downstream, left, and right of the midpoint of the fish's current cell.) However, there may be only three (or even only two) adjacent cells if the fish's current cell is on the first or last transect of a reach, or is the first or last cell on its transect. Cells from other reaches are not included among the adjacent cells always included as potential destinations.

Barriers—Barriers (section 5.1.1) can affect which cells are potential movement destinations. For cells upstream of a trout's current cell, the effect is straightforward: if there is a barrier between the trout and an upstream cell, then the upstream cell is excluded as a potential destination. Barrier locations are defined by an X coordinate, and barriers usually occur at the border between two transects. For determining whether a barrier is upstream or downstream of a fish, the location of the barrier is compared to the X coordinate of the midpoint of the cell occupied by the fish.

For cells downstream of a trout's current cell, the barrier effect is less straightforward. Fish are assumed to consider as movement destinations those cells they are familiar with, as determined by *maxMoveDistance*. However, it is not reasonable to assume that a fish is familiar with habitat downstream of a barrier, although real fish do sometimes move downstream over barriers. Such "blind" movements seem most likely to be adaptive if they are undertaken only when the alternative of not moving downstream is clearly bad.

Some literature shows that trout will move irreversibly into unknown habitat when expected growth conditions are low but before they actually reach poor condition. Wilzbach (1985) observed that trout left a stream channel soon after food availability was severely reduced, long before starvation would have been a risk. Keeley (2001) found no statistically significant difference in the condition of steelhead juveniles that emigrated from an overpopulated experimental channel versus those that remained.

Downstream movement over barriers is modeled this way:

- Cells within *maxMoveDistance* from a fish, but downstream of a barrier, are included as potential movement destinations.

- Fish are assumed unfamiliar with the actual conditions at cells downstream of a barrier, and therefore cannot predict the "expected maturity" fitness measure used to evaluate destinations (section 6.2.3). Therefore, fish are assumed to move downstream over a barrier only if none of the alternative cells (which do not require crossing the barrier) have an expected maturity value greater than the parameter *fishEMForUnknownCells*. In the absence of conclusive information on when trout will cross downstream barriers, *fishEMForUnknownCells* is arbitrarily given a value of 0.1. Therefore, a fish will not cross a barrier unless its expected survival at cells above the barrier is low (the expected probability of surviving over the fitness time horizon, *fishFitnessHorizon*, is less than *fishEMForUnknownCells*). But, when expected survival above the barrier is low, the fish always crosses the barrier. For example, if feeding conditions above a barrier become poor, fish will remain above the barrier until their condition drops enough that starvation is a high risk before moving downstream over the barrier. [This assumption has not yet been thoroughly explored or tested; Gowan and Fausch (2002) presented a more elaborate way to estimate fitness benefits of unknown locations.]

- If a fish crosses a downstream barrier, it is placed in a cell just across the barrier. Then the fish immediately repeats its entire habitat selection method, allowing it to find a good cell below the barrier. This assumption means a fish can cross multiple barriers in a day.

Minimum depth—Cells are excluded as destinations if they have depth ≤ 0. This criterion is imposed to reduce computer execution times: although the fitness measure fish use to evaluate potential destinations (section 6.2.3) could be applied to dry cells, specifically excluding movement to dry cells significantly reduces the computations needed to select a destination cell.

Fish are not required to move out of their current cell if its depth drops to zero, but they will have a strong incentive to move. However, if the flow decreases so that the nearest cell with nonzero depth is farther away than a fish's maximum movement distance (not unlikely for very small fish), then the fish can be trapped in a dry cell. (See section 6.4.3 concerning stranding mortality.)

6.2.3. Evaluate potential destination cells—

A fish evaluates each potential destination cell to determine the fitness it would provide, using the "expected maturity" fitness measure of Railsback et al. (1999). Each fish has to evaluate its potential fitness in each potential destination cell (and its current cell), because expected maturity is a function of the fish's size and condition and the cell characteristics.

Individual fish select the potential destination cell providing the highest value of *expectedMaturity* where:

$$expectedMaturity = nonStarvSurvival \times starvSurvival \times fracMature.$$

The variable *nonstarvSurvival* is the calculated probability of survival for all mortality sources except poor condition, over a specified time horizon given by the parameter *fishFitnessHorizon*. This method assumes that fish use a simple prediction of future survival: over the time horizon, the daily survival probability for risks other than poor condition within a cell equals the current day's risks. The value of *nonStarvSurvival* is calculated as:

$$nonStarvSurvival = (S_i \times S_{ii} \times S_{iii...})^{fishFitnessHorizon}$$

where S_i, S_{ii}, S_{iii}, etc. are the daily survival probabilities for all the mortality sources (*i, ii, iii, …*), evaluated for the current day, fish, and cell (these probabilities are described in section 6.4). To simplify the model's software, the value of *nonStarvSurvival* is determined for the fish's size **before** the daily growth that would occur at the potential destination cell.

The formulation of *nonStarvSurvival* assumes that trout consider all mortality sources in their habitat selection decision. This means that the trout are assumed to be aware of all the kinds of mortality in the model and are able to estimate the risk posed by each. This assumption seems reasonable for all the mortality sources currently in InSTREAM, but may not be appropriate for some new kinds of mortality that could be added to InSTREAM. In particular, angler harvest is a source of mortality that, if added to InSTREAM, might best be represented without including it in *nonStarvSurvival*—i.e., by assuming that trout cannot sense or base decisions on the risk of angler harvest.

In the equation for *expectedMaturity*, the value of *starvSurvival* is the probability of surviving the risk of poor condition (closely related to starvation; section 6.4.4) over the number of days specified by the parameter *fishFitness-Horizon*. This term introduces the effects of food intake to the fitness measure. The method assumes that fish evaluate *expectedMaturity* using the simple prediction that the current day's growth rate will persist over the time horizon. The value of *starvSurvival* is determined by the following steps (Railsback et al. 1999):

- Determine the foraging strategy, food intake, and growth (g/d) for the fish and habitat cell in question, for the current day, using the methods in section 6.3.

- Project the fish's weight, length, and condition factor *fishCondition* (section 6.3.1) that would result if the current day's growth persisted over the fitness time horizon specified by *fishFitnessHorizon*. The daily growth is multiplied by *fishFitnessHorizon* to determine the change in weight over the time horizon; the corresponding change in length and condition factor, *K* (weight to length ratio) are determined using methods in section 6.3.1.

- Approximate the probability of surviving starvation over the fitness horizon, estimated as the first moment of the logistic function of poor condition survival versus *K* (section 6.4.4):

$$starvSurvival \; = \; \left[\frac{1}{a} \ln \left(\frac{1 + e^{(a\,K_{t+T} + b)}}{1 + e^{(a\,K_t + b)}} \right) \middle/ (K_{t+T} - K_t) \right]^{T}$$

where K_t is the fish's value of *fishCondition* at the current day and K_{t+T} is the projected condition factor at the end of the fitness horizon, T is equal to *fishFitnessHorizon*, and a and b are the *logistA* and *logistB* variables (determined within the code from parameter values; see the logistic function conventions described in chapter 1) for poor condition mortality. This equation would cause a divide-by-zero error when K_{t+T} equals K_t, a common condition because *K* equals 1.0 whenever fish are well fed. This equation is also subject to significant errors owing to the limits of computer precision when K_{t+T} is extremely close to K_t. To avoid these problems, *starvSurvival* is set equal to the daily survival probability for K_t, raised to the power *fishFitnessHorizon*, whenever the difference between K_{t+T} and K_t is less than 0.001.

The final term in the equation for *expectedMaturity* is *fracMature*, which represents how close to the size of sexual maturity a fish would be at the end of the fitness time horizon. It is simply (a) the length the fish is projected to be at the end of the time horizon, divided by (b) the parameter *fishSpawnMinLength* (section 6.1.1), and limited to a maximum of 1.0. This term gives juvenile trout an incentive to select cells with higher growth, encouraging them to reach reproductive maturity.

The time horizon variable *fishFitnessHorizon* is the number of days over which the terms of the expected maturity fitness measure equation are evaluated. The biological meaning of this variable is the time horizon over which fish evaluate the tradeoffs between food intake and mortality risks to maximize their probability of surviving and reproducing. It is discussed in the "unified foraging theory" (also called "dynamic state variable modeling" literature: Clark and Mangel 2000, Houston and McNamara 1999, Mangel and Clark 1986). Longer time horizons better reflect how an individual's fitness depends on how well it makes decisions throughout its reproductive life. However, the simple prediction used to evaluate *expectedMaturity*—that habitat and competitive conditions are constant over the time horizon—becomes very questionable for long time horizons. Smaller values of *fishFitnessHorizon* place less emphasis on food intake and avoiding starvation in movement decisions. Values of *fishFitnessHorizon* of 5 to 10 d cause *expectedMaturity* to vary almost exclusively with nonstarvation survival. Values of *fishFitnessHorizon* in the range of 100 d cause *expectedMaturity* to vary almost exclusively with growth rates when growth is less than the minimum needed to maintain a condition factor of 1.0.

A few studies have addressed the issue of fitness time horizons. Bull et al. (1996) used a decisionmaking model similar to habitat selection in InSTREAM and assumed overwintering juvenile salmon used the remaining winter period as a time horizon. Thorpe et al. (1998) proposed using the duration of various salmonid life stages as time horizons. Assuming that fish anticipate seasonal changes in habitat conditions and their life stage, it makes sense to assume they use a habitat selection time horizon of several months. An analysis of the sensitivity of InSTREAM to the value of *fishFitnessHorizon* (discussed in chapter 3) shows that population success, evaluated as mean adult trout biomass, was relatively insensitive to the value of *fishFitnessHorizon*, and relatively high over a range of about 70 to 120 d (fig. 8). Most applications of InSTREAM to date have used a value of 90 d.

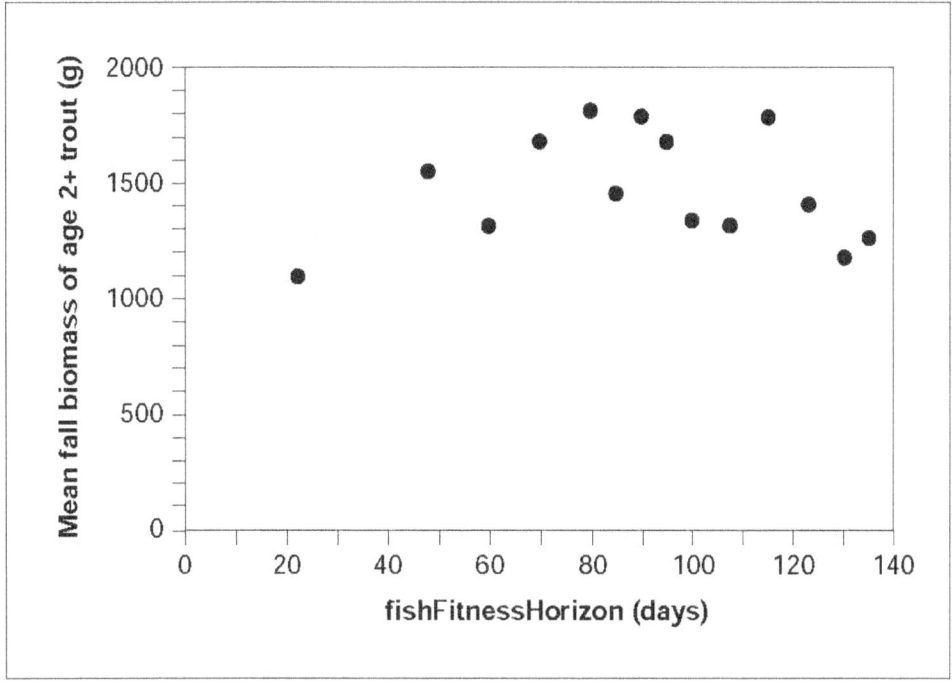

Figure 8—Sensitivity of trout biomass to the value of *fishFitnessHorizon*. The Y axis is the average September biomass of adult (age 2 and older) trout, over 11 simulated years at the Little Jones Creek lower study site.

6.2.4. Move to best destination—

The fish identifies the cell that has the highest value of the expected maturity fitness measure, and then moves there. When a fish moves into a cell, the resources it uses are subtracted from those available for subsequent fish (sections 5.2.3, 5.2.6). These resources may include one of the two kinds of food, and velocity shelter. A fish may move into a cell even when none of these resources remain available to it, in which case its consumption of them is zero.

6.3. Feeding and Growth

6.3.1. Overview—

Methods for determining the daily growth—change in weight and length—are used both in the habitat selection decision and to simulate growth (the third daily action by fish, section 12.2). This first subsection provides an overview of the feeding and growth methods, listing the major assumptions. Full detail is provided starting with section 6.3.2.

The feeding and growth formulation of InSTREAM is conceptually related to a number of other models. It borrows both basic concepts and detailed methods from previous research on fish energetics. The concepts of modeling growth as net energy intake (the difference between energy input from food and energy consumption for metabolism) and modeling metabolic energy consumption as a function of fish size, swimming speed, and temperature are well-established in the literature (Hanson et al. 1997; see also Brandt and Hartman 1993, Elliott and Hurley 2000). Similar to several previous efforts (e.g., Braaten et al. 1997, Fausch 1984, Gowan and Fausch 2002, Grossman et al. 2002, Hayes et al. 2000, Hill and Grossman 1993, Hughes and Dill 1990, Van Winkle et al. 1998), inSTREAM combines bioenergetics models and feeding models to predict net energy intake as a function of fish size and habitat conditions (especially, depth and velocity).

One important way that InSTREAM differs from previous feeding and growth models is its inclusion of competition among individual fish for food. A fish's food intake is assumed to be limited by either the availability of food or the ability of the fish to capture food. The ability to capture food depends on fish size (increasing with length, because larger fish see and swim better) and on habitat conditions such as velocity and depth. Food availability depends on how much food is produced in the cell and how much is consumed by competing fish (section 5.2.6).

Fish in InSTREAM feed during daylight hours and never at night—a major simplifying assumption. Recent literature shows that night feeding is not unusual and under some conditions is more common than daytime feeding (e.g., Bradford and Higgins 2001, Fraser and Metcalfe 1997, Metcalfe et al. 1999). Whether an individual trout feeds during day or night (or neither) appears to emerge from how mortality risk and food intake vary between day and night, which can in turn vary with fish size, competition, and many habitat variables. InSTREAM has, in fact, been modified (version 3, not described here) to simulate how trout choose between feeding during day and night (Railsback et al. 2005), but this capability requires a major increase in the model's complexity. Although the assumption that trout feed during daytime only is clearly not always realistic, it is often useful for the intended purposes of InSTREAM.

A related simplifying assumption is that trout do not feed at temperatures below a threshold specified by *fishMinFeedTemp* (provisionally 2 °C). In reality, trout have been observed to feed less, and more nocturnally, as temperature decreases toward freezing (Cunjak et al. 1998, Fraser et al. 1993). The use of a threshold temperature is a simplification of this process. Hill and Grossman (1993) observed a rapid drop in feeding activity in rainbow trout at 2 °C.

InSTREAM does not specify the exact kinds of food consumed by fish, but its feeding formulation and parameters generally represent invertebrate food. Even though the model assumes small fish are vulnerable to predation by adult trout (section 6.4.6), fish generally do not make up a large part of the diet of stream trout. Therefore, piscivory is not represented in the feeding methods.

Fish in InSTREAM can use either of two feeding strategies. Drift feeding, in which the fish remains stationary and captures food as it is carried past by the current, is often the most profitable strategy (Fausch 1984, Hill and Grossman 1993, Hughes and Dill 1990). Drift food intake is modeled as a function of stream depth and velocity and fish length; intake peaks at an optimal velocity that increases with fish size. Drift intake decreases as turbidity increases and makes it harder for fish to detect food items. Metabolic costs for drift feeding increase with water velocity, but use of velocity shelters reduces this cost. The second feeding strategy is active searching for food. Search feeding can be important when competition for food is intense, conditions for drift feeding are poor, or benthic food is abundant (Fausch et al. 1997, Nielsen 1992, Nislow et al. 1998). The energetic benefits of search feeding are assumed to be mainly a function of food availability, with energetic cost depending on water velocity.

The feeding and growth methods calculate the potential food intake and metabolic costs a fish would experience in a cell, for both drift and search feeding. Standard bioenergetics approaches (Hanson et al. 1997) are used by InSTREAM to calculate net energy intake for each feeding strategy. Net energy (the difference between energy intake from food and metabolic energy cost) is often negative. The fish then selects the strategy that provides the highest net energy intake. Growth is proportional to net energy intake.

A fish's length and condition factor are updated from its daily growth. How an organism allocates its energy intake to growth (increase in length), storage (increase in weight or fat reserves), or gonads is in reality a complex, adaptive process. For example, a juvenile fish may reduce its risk of predation most by increasing in length as rapidly as possible, but allocating all energy intake to growth instead of storage increases the risk of starvation during periods of reduced intake. However, InSTREAM does not model energy allocation as an adaptive trait. Instead it uses the approach of Van Winkle et al. (1996) that simply forces fish to maintain a standard relation between length and weight during periods of positive growth.

The method for calculating daily change in length adopted from Van Winkle et al. (1996) uses their nonstandard definition of a condition factor. In fisheries science, a condition factor is a unitless index of a fish's weight relative to its length. A higher condition factor indicates that a fish is heavy for its length and has high energy reserves, and therefore less vulnerable to starvation or disease during periods of negative growth. The condition factor index used in InSTREAM (*fishCondition*, not to be confused with the more traditional K) can be considered the fraction of "healthy" weight a fish is, given its length. The value of *fishCondition* is 1.0 when a fish has a "healthy" weight for its length, according to a length-weight relation input to the model via fish parameters *fishWeightParamA* and *fishWeightParamB*:

$$fishHealthyWeight = fishWeightParamA \times fishLength^{fishWeightParamB}$$

Fish grow in length whenever they gain weight while their value of *fishCondition* is 1.0. Condition factors less than 1.0 indicate that the fish has lost weight. In this formulation, values of *fishCondition* cannot be greater than 1.0. Weight (*fishWeight*, g), length (*fishLength*, cm), and *fishCondition* are calculated in this way.

- The fish's new weight is determined by adding its daily growth (which can be negative) to its previous weight.
- The fish's new weight is used, with the inverted length-weight relation for healthy fish, to calculate *fishWannabeLength*, the length the fish would be if its condition factor were 1.0:

$$fishWannabeLength = \left(\frac{fishWeight}{fishWeightParamA} \right)^{1/fishWeightParamB}$$

- If the fish's current length is less than *fishWannabeLength* (indicating that the fish is not underweight), then its new length is set to *fishWannabeLength*. The fish grows in length while keeping its *fishCondition* value equal to 1.0.
- If the fish's current length is greater than *fishWannabeLength* (indicating that the fish is underweight for its length), its length remains unchanged.
- The new value of *fishCondition* is equal to the fish's new weight divided by the "healthy" weight for a fish its length:

$$fishCondition = fishWeight \Big/ \left(fishWeightParamA \times fishLength^{fishWeightParamB} \right)$$

This formulation is simple and succeeds in producing reasonably realistic patterns of trout growth under many conditions. However, the formulation has several noteworthy limitations:

- Fish cannot store energy reserves. Fish will have a condition of 1.0 **only** on those days when growth is positive. Even if a fish has eaten well for many days in succession, its *fishCondition* can only be as high as 1.0, and 1 day of negative net energy intake causes condition to fall below 1.0. This could be important under conditions of highly variable food intake because survival is assumed to decrease with condition (section 6.4.4).

- This weight-based condition factor is not the best predictor of starvation mortality (section 6.4.4).

- Calibration of growth to situations where this relationship is valid will be automatic, but calibration to situations where the relationship is not valid will be impossible. For example, InSTREAM cannot predict the existence of unusually fat fish.

- The energetics of reproduction are not considered. Although InSTREAM simulates weight loss from spawning (section 6.1.5), it models neither storage of energy for gonad development nor how gonad production affects length and weight.

These limitations could be eliminated only by making InSTREAM considerably more complex. Methods for representing energy allocation more realistically in IBMs have not yet been developed and tested. The current formulation appears adequate and appropriate for InSTREAM's objectives.

Example parameter values for the length-weight relationship are provided in table 8. These parameters should not simply be regression parameters calculated from observed data; they must describe a site-specific length-weight relation for fish **in good condition**. Chapter 3 covers methods for developing the parameters.

Table 8—Example parameter values for the length (cm)–weight (g) relation

Species and site	Parameter	Value
Cutthroat trout, Little Jones Creek, Del Norte County, California[a]	fishWeightParam A	0.0124
	fishWeightParam B	2.98
Rainbow trout, Tule River, Tulare County, California[b]	fishWeightParam A	0.0134
	fishWeightParam B	2.96
Brown trout, Tule River, Tulare County, California[b]	fishWeightParam A	0.0123
	fishWeightParam B	2.97

[a] Source: Railsback and Harvey (2001)
[b] Source: Van Winkle et al. (1996)

6.3.2. Activity budget—

Energy intake and costs differ between feeding and resting fish. Energetic calculations are based on hourly energy rates (J/h), and the daily energy totals depend on how many hours are spent feeding versus resting. Trout are assumed to spend all daylight hours feeding and all night hours resting, except that no feeding occurs when the temperature is less than *fishMinFeedTemp*. Daylight hours are assumed to include 1 hour before sunrise and 1 hour after sunset. Consequently, the time spent feeding per day (*feedTime*, h/d) is:

$$feedTime = dayLength + 2;$$
$$\text{if } temperature < fishMinFeedTemp, \text{ then } feedTime = 0.$$

6.3.3. Food intake: drift feeding strategy—

Drift-feeding fish capture invertebrates as they are carried within range by the current. The drift-feeding energy intake formulation of InSTREAM is unique but related to the previous feeding and net energy intake models cited in section 6.3.1. This literature shows that the distance over which fish can see and capture food increases with trout size and decreases with water velocity. Unlike previous models, InSTREAM includes the negative effect of turbidity on the ability of trout to see and capture prey. Turbidity can vary dramatically among sites and over time, and its effects on trout feeding are strong and relatively predictable. Unlike some previous models of drift feeding, InSTREAM does not incorporate prey size. Prey size is naturally variable and unpredictable, and its effects could not easily be distinguished from those of other factors. This drift-feeding formulation differs from previous versions of InSTREAM and the predecessor model of Van Winkle et al. (1996; see also Van Winkle et al. 1998). The revision was made to make the best use of available literature.

Drift-feeding fish are assumed to capture some of the food items that pass within a "capture area" (*captureArea*, cm^2), a rectangular area perpendicular to the current, the dimensions of which depend only on fish size (explained below). The fraction of food items passing through the capture area that are actually caught (*captureSuccess*, unitless) decreases with cell velocity, increases with fish swimming ability, and decreases with turbidity. A fish's intake rate (*driftIntake*, g/h) is calculated as the mass of prey passing through the capture area times the capture success:

$$driftIntake = captureSuccess \times habDriftConc$$
$$\times velocity \times captureArea \times 3,600 \text{ s/h.}$$

In this equation, *habDriftConc* (g/cm^3) is a habitat reach variable (section 5.1.1).

A detection distance approach is used to calculate *captureArea*. Detection distance is defined as the distance over which fish can see and attack—but not necessarily capture—prey. Detection distance is believed to depend primarily on the size of the fish (bigger fish have bigger, more sensitive eyes) and the size of the prey (bigger prey being easier to detect). Schmidt and O'Brien (1982) collected empirical data on how detection distance in a stream salmonid (arctic grayling [*Thymallus arcticus*]) varied with fish and prey size. These experiments used zooplankton as prey, but their results have been used successfully as the basis of drift feeding models by Hughes (1992b) and Hughes et al. (2003). Schmidt and O'Brien (1982) measured detection distance of fish with lengths from 3 to 13 cm during daylight and night conditions and for a variety of zooplankton prey sizes. Only daylight observations for 0.2-cm prey (the largest) are used here.

These observations can be represented with a linear model of detection distance versus fish length using the data of Schmidt and O'Brien (1982) (fig. 9). A logarithmic equation fits the data more closely, but the linear model was chosen for several reasons. First, it captures the fact that very small trout cannot use as wide a range of prey sizes as larger trout can, which is not otherwise represented in the feeding model. Second, a logarithmic fit to these data predicts negative detection distances for trout lengths less than 2 cm and does not reproduce the observations of Hughes et al. (2003) that detection distance continues to increase to over 100 cm for very large trout. Finally, precalibration of the growth model was used to select the intercept and slope of the linear model (parameters *fishDetectDistParamA* and *fishDetectDistParamB* defined below). The precalibration analysis indicated that the growth rates of very small trout were very sensitive to the intercept. An intercept of 4.0 was found to provide growth of very small trout that was realistic at the same drift-food availability values that produce realistic growth rates in larger trout.

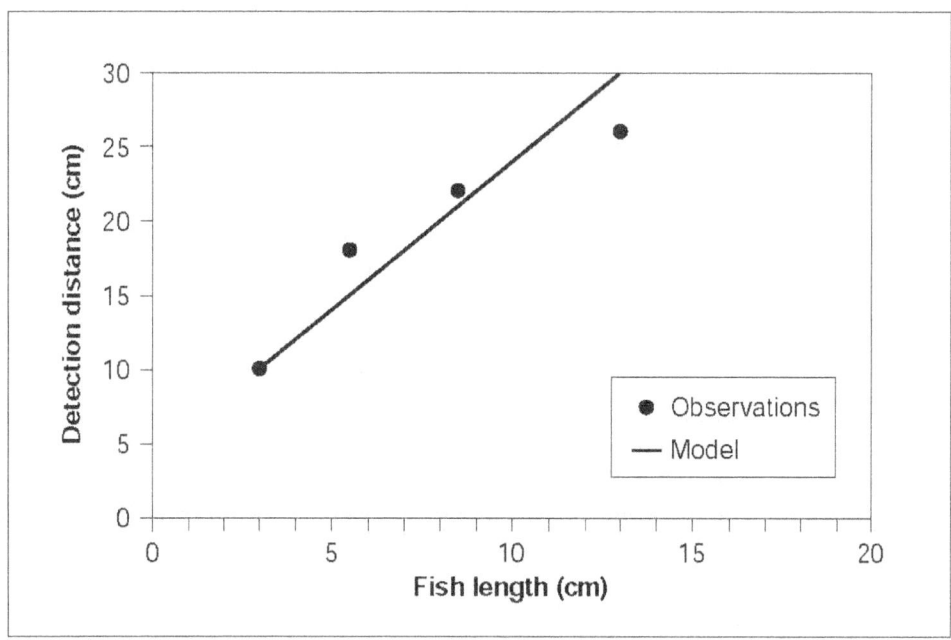

Figure 9—Relation between fish length and prey detection distance observed by Schmidt and O'Brien (1982), for arctic grayling feeding on 0.2-cm zooplankton.

Detection distance is adjusted for turbidity. The primary effect of turbidity on drift feeding is to reduce the ability of fish to detect prey: Sweka and Hartman (2001) observed that as turbidity increased (in the range 3 to 40 NTU) the frequency of prey detection by trout decreased, but the frequency of attacking and capturing detected prey did not decrease. Sweka and Hartman (2001) developed a curve for how detection distance decreases with turbidity, for 14-cm brook trout (*Salvelinus fontinalis*) feeding on large (1.0 cm), floating prey. The function used by InSTREAM for **relative** detection distance (the fractional reduction in detection distance at turbidity levels above zero) is based on the data of Sweka and Hartman (2001) but differs from their curve in two ways.

First, InSTREAM assumes that turbidity has no effect at values below a threshold of 5 NTUs (defined by the parameter *fishTurbidThreshold*). The curve of Sweka and Hartman (2001) has a steep gradient at low turbidity levels, which would make feeding success very sensitive to low turbidity values. However, none of the available data (Barrett et al. 1992, Sweka and Hartman 2001) clearly show an effect of turbidity at levels below 5 NTUs (fig. 10), and it seems likely that below such a threshold, reactive distance is limited by other factors such as turbulence and the ability (or net benefit) of catching food items very far away. Another reason for assuming a turbidity threshold is to avoid making InSTREAM highly sensitive to low turbidity levels, which are hard to measure or estimate accurately.

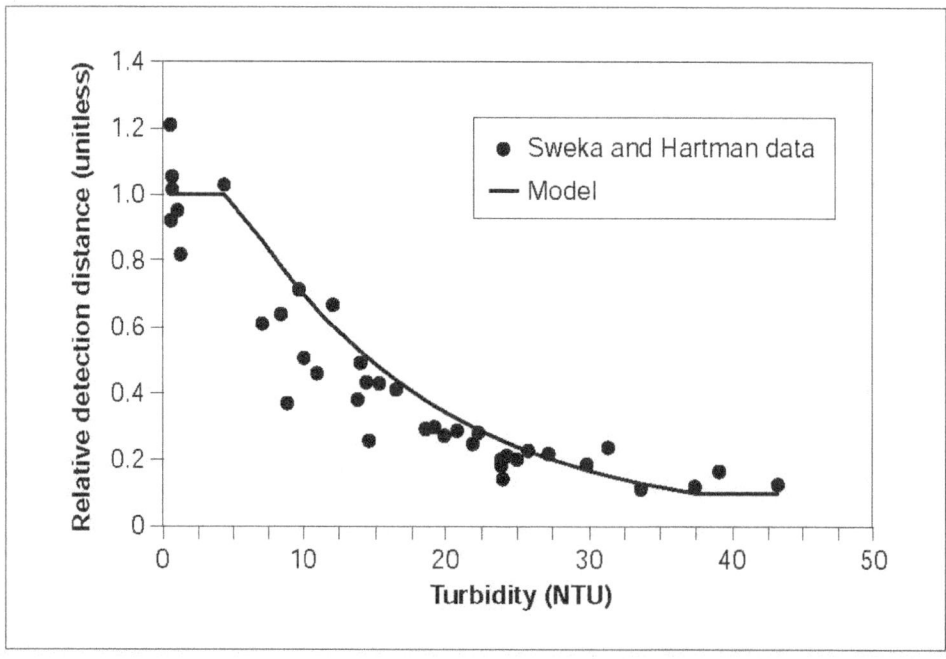

Figure 10—Relative detection distance vs. turbidity: model and data of Sweka and Hartman (2001) used to fit the model.

The second change adds a minimum detection distance. Sweka and Hartman's (2001) observations indicate that detection distance does not go completely to zero as turbidity exceeds 50 NTUs. Harvey and White (2008) provided support for this approach with observations that both cutthroat trout and coho salmon captured some drift in turbidity of 100 NTU. Therefore, InSTREAM includes a parameter *fishTurbidMin* which limits the effect of turbidity on detection distance (fig. 10).

Detection distance is therefore modeled with this equation:

$$detectDistance = [fishDetectDistParamA + (fishDetectDistParamB \times fishLength)] \times turbidityFunction$$

where:

if $habTurbidity \leq fishTurbidThreshold$ then $turbidityFunction = 1.0$
else $turbidityFunction = \max[\exp(fishTurbidExp \times (habTurbidity - fishTurbidThreshold)), fishTurbidMin]$.

Table 9 presents the parameter values and figure 10 the resulting model. The value of *fishTurbidExp* was fit via regression to the data of Sweka and Hartman (2001) by (1) establishing the reactive distance for negligible turbidity as the mean of reactive distances observed at turbidities less than 5 NTU (the seven such observations had a mean reactive distance of 80.8 cm); (2) calculating the relative reactive distance for other observations as the observed reactive distance divided by 80.8; and (3) using exponential regression on relative reactive distance versus (turbidity – 5 NTU) (the regression line was forced through the point (0,1), so relative reactive distance is 1 when turbidity is 5).

Table 9—Detection distance and capture success parameters

Parameter	Definition	Value
fishDetectDistParamA	Intercept in equation for detection distance (cm)	4.0
fishDetectDistParamB	Multiplier in equation for detection distance (unitless)	2.0
fishTurbidThreshold	Highest turbidity that causes no reduction in detection distance (NTU)	5.0
fishTurbidExp	Multiplier in exponential term for the turbidity function (unitless)	-0.0711
fishTurbidMin	Minimum value of the turbidity function (unitless)	0.1
fishCaptureParam1	Ratio of cell velocity to fish's maximum swim speed at which capture success is 0.1 (unitless)	1.6
fishCaptureParam9	Ratio of cell velocity to fish's maximum swim speed at which capture success is 0.9 (unitless)	0.5

NTU = nephelometric turbidity units.

Several previous trout-feeding models assumed that the capture area is a circle or half-circle with radius equal to the detection distance, but Booker et al. (2004) showed that failing to consider depth (which often is less than the detection distance) can cause major errors. InSTREAM uses a capture area for drift feeding that depends on the detection distance and cell depth. The width of the rectangular capture area is twice the detection distance: fish are assumed to detect all drift that comes within the detection distance to their left and right, as they face into the current. The height of the capture area is the minimum of the reactive distance and the depth, as fish are assumed more likely to be near the stream bottom than at middepth when feeding:

$$captureArea = [2 \times detectDistance] \times [\min(detectDistance, cellDepth)].$$

Whereas the capture area models the area over which drift-feeding trout can detect prey, **capture success** models what fraction of detected prey are actually caught. Capture success is largely a function of water velocity. Fish must be able to swim to the prey, capture it, and return to their feeding station. At higher velocities, maneuvering quickly enough to capture prey is more difficult, and swimming longer distances after prey requires more energy (because the fish must swim back upstream to return to their feeding station; Hughes et al. 2003). Capture success is also affected by temperature, as the ability of fish to maneuver and swim rapidly is reduced at low temperatures.

Hill and Grossman (1993) measured capture success for rainbow trout feeding on 0.2-cm prey. The trout had lengths of 6 and 10 cm, and measurements were made at 5 and 15 °C with velocities ranging from 0 to 40 cm/s. Capture success was evaluated as the fraction of prey within the fish's detection distance that was caught. Hill and Grossman (1993) approximated the detection distance as 2.5 times the fish's standard length, which is fairly close to the detection distance used in InSTREAM (fig. 9). Hill and Grossman measured capture success within each of three ranges: the inner 20 percent of the capture distance, 20 to 60 percent of capture distance, and 60 to 100 percent of capture distance. To develop parameters for InSTREAM, these values were averaged over the entire capture distance. For all the observations (35 combinations of fish size, temperature, and water velocity), capture success fit a logistic function of the ratio of water velocity to maximum sustainable swimming speed of the fish (fig. 11):

$$captureSuccess = \text{logistic}(habVelocity / fishMaxSwimSpeed).$$

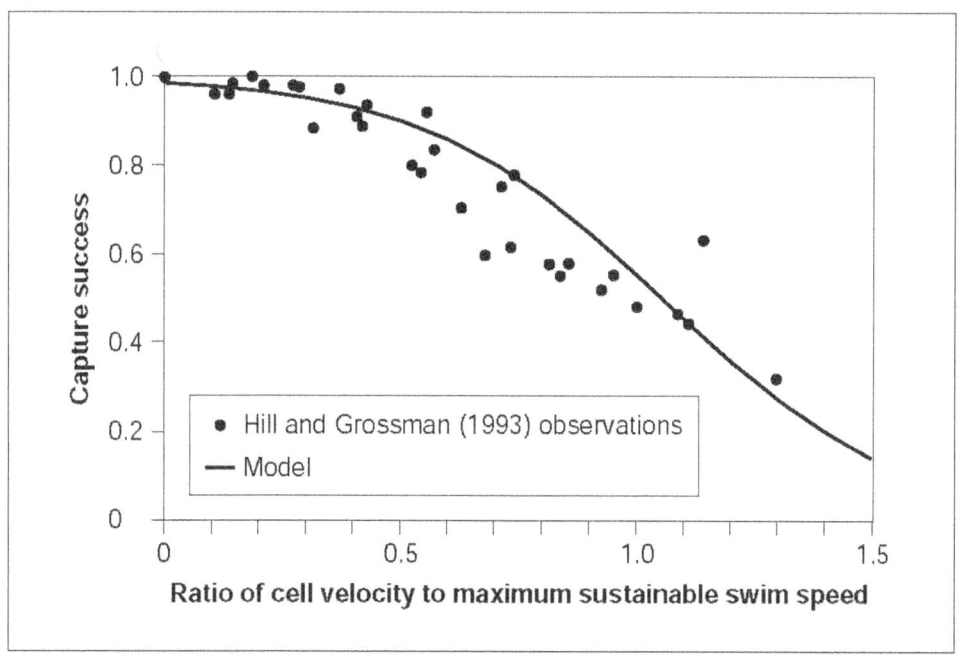

Figure 11—Capture success model and the Hill and Grossman (1993) observations it was based on.

Maximum sustainable swimming speed (*fishMaxSwimSpeed*) is a function of fish length and water temperature (section 6.4.2.). Maximum sustainable swim speed appears to be useful for modeling capture success for two reasons. First, it scales capture success with both fish length and temperature. Second, Hughes et al. (2003) observed that large brown trout actually swim at sustainable (or even lower) speeds when capturing food. Parameters for this logistic function are in table 9.

Sensitivity of the drift-feeding model to velocity and fish size is explored in section 6.3.10.

6.3.4. Food intake: active searching strategy—

Actively searching for benthic or drop-in food is an alternative to the drift-feeding strategy, but no established models address search feeding by trout. An optimal foraging approach would be to assume fish search for food at a rate that maximizes the difference between energy intake from feeding and energy cost of swimming. To avoid the complexity of such an approach, InSTREAM simply assumes that the rate of search-food intake is proportional to the rate at which search food becomes available: every fish searches for food at about the same rate, so intake increases linearly with food production. We assume search-feeding intake decreases linearly to zero as water velocity increases to the fish's maximum sustainable swim speed. This velocity function represents how the ability of a fish to see and search for food decreases with velocity. (It does **not** represent the energetic cost of swimming at high velocities, which is considered in the respiration formulation, section 6.3.7.)

The search food intake model is:

$$searchIntake = habSearchProd \times fishSearchArea$$
$$\times \max\left(\left[\frac{fishMaxSwimSpeed - cellVelocity}{fishMaxSwimSpeed}\right], 0\right)$$

where *searchIntake* (g/h) is the rate at which food is taken in via search feeding, *habSearchProd* (g wet weight/h-cm^2) is the search food production rate (section 5.2.6), *fishMaxSwimSpeed* is the fish's maximum sustainable swimming speed (cm/s; section 6.4.2), and *cellVelocity* (cm/s) is the water velocity in the fish's cell. The proportionality constant *fishSearchArea* (cm^2) can be loosely interpreted as the area over which the production of stationary (nondrifting) food is consumed by one fish. This search area is not necessarily a contiguous piece of stream area: a small fish searching a small area closely may obtain the same food intake as a big fish spot-searching over a much larger area. Because *habSearchProd* and *fishSearchArea* have the same effect on search intake and both would be very difficult to measure, either would be a good parameter to use for calibration. Note that fish size does not affect search food intake except for the effect of size on *fishMaxSwimSpeed*. Therefore, search feeding is more likely to be the desirable strategy for smaller fish.

Note that turbidity does not affect search feeding in the model. Although search feeding can sometimes be primarily visual, observations of similar stomach fullness for trout within sites across a range of turbidity conditions suggest that salmonids can search-feed successfully using other senses (e.g., White and Harvey 2007). DeRobertis et al. (2003) conducted tank experiments on feeding by juvenile chum salmon (*Oncorhynchus keta*) at various turbidity levels, using planktonic prey in standing water. Feeding success under daytime conditions did not decrease consistently at turbidities between zero and 20 NTU; at 40 NTU, feeding success was about one-third of that in clear water. During nighttime light levels, even turbidities up to 40 NTU, caused no decrease in feeding success. Because the effects of turbidity on search feeding are apparently limited, they are ignored in InSTREAM.

6.3.5. Food intake: maximum consumption—

As part of the net energy intake calculations, calculated food intake from drift or search feeding is checked to make sure it does not exceed the physiological maximum daily intake. This maximum daily consumption, referred to as *cMax* (g/d), represents the maximum rate of food consumption if a fish is limited only by its physiology. Field bioenergetics studies (Preall and Ringler 1989, Railsback and Rose 1999) indicated that actual food intake does not approach *cMax* under typical conditions. However, here *cMax* serves the purpose of restricting intake and growth during low temperatures, a function otherwise lacking in the model, except that the time spent feeding becomes zero below a threshold temperature (section 6.3.2). Cunjak et al. (1998) cited evidence that low food assimilation efficiencies and gut evacuation rates, which can be represented by *cMax*, limit energy intake in cold temperatures.

Unfortunately, *cMax* is poorly defined and difficult to measure, largely because it varies with factors such as the fish's exercise condition, food type, and feeding conditions (Myrick 1998, PG&E 1994). However, there are a number of published equations for *cMax* that include (a) an allometric function, relating *cMax* to fish size, and (b) a temperature function (Hanson et al. 1997). InSTREAM uses:

$$cMax = fishCmaxParamA \times fishWeight^{(1 + fishCmaxParamB)} \times cmaxTempFunction.$$

This equation is widely used with the parameters developed by Rand et al. (1993) for rainbow trout (table 10) for modeling *cMax* of salmonids in general (e.g., Booker et al. 2004, Railsback and Rose 1999, Van Winkle et al. 1996).

Table 10—Parameter values for allometric function of maximum consumption

Parameter	Definition	Value
fishCmaxParamA	Allometric constant in cMax equation (unitless)	0.628
fishCmaxParamB	Allometric exponent in cMax equation (unitless)	-0.3

The *cMax* temperature function used in InSTREAM is based in part on laboratory studies on rainbow trout by Myrick (1998) and Myrick and Cech (2000). These studies focused on higher temperatures, measuring *cMax* at 10, 14, 19, 22, and 25 °C. Previous models of *cMax* for salmonids (Rand et al. 1993) used temperature functions based on the laboratory studies of From and Rasmussen (1984), who studied rainbow trout at temperatures of 5 to 22 °C; and of Elliott (1982) who studied brown trout. InSTREAM uses values of *cmaxTempFunction* determined by interpolation from a set of seven points from these laboratory studies (table 11; fig. 12).

Table 11—Parameter values for temperature function of maximum consumption

Parameter name	Temperature	Parameter name	Temperature function value (unitless)
	Degrees Centigrade		
fishCmaxTempT1	0	fishCmaxTempF1	0.05
fishCmaxTempT2	2	fishCmaxTempF2	0.05
fishCmaxTempT3	10	fishCmaxTempF3	0.5
fishCmaxTempT4	22	fishCmaxTempF4	1.0
fishCmaxTempT5	23	fishCmaxTempF5	0.8
fishCmaxTempT6	25	fishCmaxTempF6	0
fishCmaxTempT7	100	fishCmaxTempF7	0

Note: Each row in the table defines one of the points in figure 12.

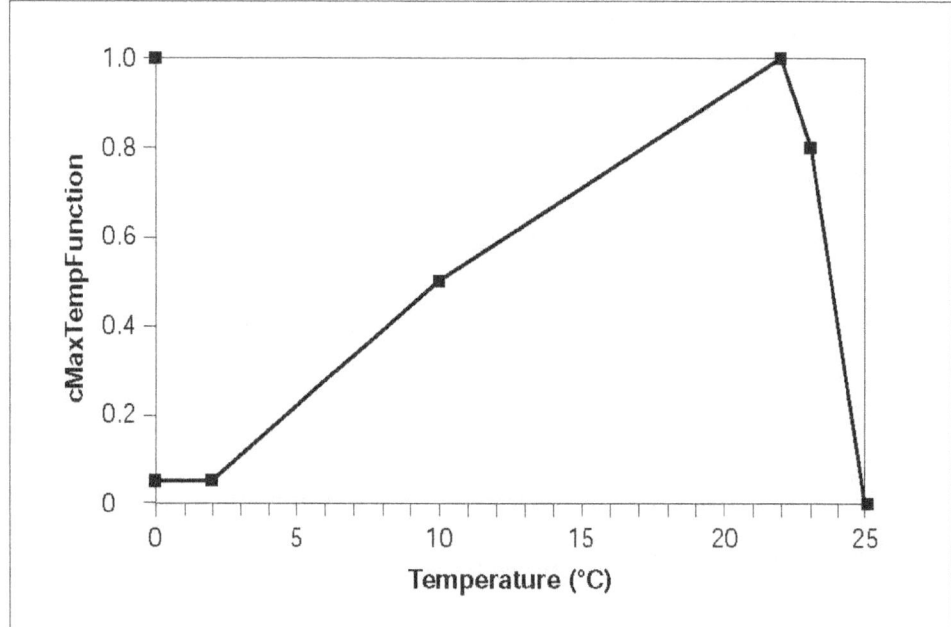

Figure 12—Temperature function for *cMax*.

Although several sets of equations and parameters for *cMax* have been published for different salmonids, careful scrutiny of these publications indicates that differences in models of *cMax* are more likely to result from differences in experimental methods than from differences among species or stocks. Considering the inherent uncertainty in *cMax* and its limited effect on results of InSTREAM, the parameters in tables 10 and 11 are cautiously recommended for all stream trout species.

6.3.6. Food intake: daily food availability—
The food intake of each fish can be limited by the total amount of drift food (*driftDailyCellTotal*, g/d) and search food (*searchDailyCellTotal*, g/d) available

each day in its cell. These daily food availability values are a function of the fish's feeding time (*feedTime*, h; section 6.3.2) because food produced during nonfeeding hours cannot be considered available to the fish. The daily food availability rates are calculated from the hourly rates described in section 5.2.6. The hourly availability rates are the rate food is produced in the cell, minus food consumption by larger fish. Therefore, hierarchical competition for food is implemented via food availability rates. Daily food availabilities are:

$$driftDailyCellAvail \ = \ driftHourlyCellAvail \ \times \ feedTime$$

and:

$$searchDailyCellAvail \ = \ searchHourlyCellAvail \ \times \ feedTime$$

where *driftHourlyCellAvail* and *searchHourlyCellAvail* are defined in section 6.3.2.

6.3.7. Respiration costs and use of velocity shelters—

Conventional bioenergetics models for fish (Hanson et al. 1997) model respiration as the energetic cost of metabolism and swimming. This approach is adopted for InSTREAM, modeling (a) standard respiration that is independent of the fish's activity and (b) an additional cost that increases with swim speed.

Swim speeds—Drift-feeding fish are assumed to swim at their habitat cell's water velocity unless they have access to velocity shelter. Fish using the search-feeding strategy are assumed to swim at a speed equal to their cell's water velocity. These two assumptions are a highly simplified representation of how real trout swim, but the consequent error in respiration costs is neglected to simplify the model.

If a drift-feeding fish has access to velocity shelter, then its *swimSpeed* equals a constant fraction of its habitat cell's mean water velocity. This fraction is defined by the parameter *fishShelterSpeedFrac*. A number of studies have shown that "focal" water velocities (the velocity measured as closely as possible to the spot where a fish was drift-feeding) are related to, but less than, the depth-averaged velocity at the same location (e.g., Baltz and Moyle 1984, Baltz et al. 1987, Moyle and Baltz 1985). However, relations between focal and depth-averaged velocities observed in these studies are not directly applicable to InSTREAM because *fishShelterSpeed-Frac* approximates the difference between **cell average** water velocity and the swimming speed of a fish using velocity shelter. The best value of this parameter will vary with the kind of velocity shelter being used. For a small, hydraulically complex stream with velocity shelter from boulders and logs, Railsback and Harvey (2001) used a value of 0.3 for *fishShelterSpeedFrac*. An application of InSTREAM to the Green River, Utah (Railsback et al. 2005), where substrates are relatively small and embedded, used a value of 0.5.

Velocity shelter access—Model trout are assumed to compete for velocity shelter space like they compete for food. The following rules determine whether each fish has access to shelter in a habitat cell.

- Each cell has a limited area of velocity shelter; this area differs among cells but remains constant over time (section 5.2.3).
- Each drift-feeding fish is assumed to use an area of velocity shelter equal to the square of its length.
- A fish has access to velocity shelter in a cell only if the sum of shelter areas occupied by larger drift-feeding fish in the cell is less than the cell's total shelter area.

Each fish is assumed to use only a small shelter area (the square of its length) to ensure that fish compete with each other for food, not for shelter area, unless velocity shelter clearly limits net energy intake.

Respiration cost model—InSTREAM uses the Wisconsin Model equation 1 for respiration (Hanson et al. 1997), as modified by Van Winkle et al. (1996) to apply the activity respiration rate only during active feeding hours. The parameters that Rand et al. (1993) developed for steelhead trout (converted from calories to joules; table 12) are widely used and appear to be the best available for stream trout in

Table 12—Parameter values for respiration

Parameter	Definition	Units	Value
fishRespParamA	Allometric constant in standard respiration equation	—	30.0
fishRespParamB	Allometric exponent in standard respiration equation	none	0.784
fishRespParamC	Temperature coefficient in standard respiration equation	1/°C	0.0693
fishRespParamD	Velocity coefficient in activity respiration equation	s/cm	0.03

— = Empirical parameter with units that depend on *fishRespParamB*.

general. This formulation breaks respiration into two parts: standard respiration (*respStandard*, J/d) takes place 24 h/d and includes no effect of activity; activity respiration (*respActivity*, J/d) is the energy needed to swim during feeding. Total respiration (*respTotal*, J/d) is the sum of these two. The equations are:

$$respTotal = respStandard + respActivity,$$
$$respStandard = (fishRespParamA \times fishWeight^{fishRespParamB})$$
$$\times \exp(fishRespParamC \times temperature),$$

and

$$respActivity = \left(\frac{feedTime}{24}\right) \times \left[\exp(fishRespParamD \times swimSpeed) - 1\right] \times respStandard$$

Data collected by Myrick (1998; see also Myrick and Cech 2000) indicate that the standard respiration formulation overestimates the effect of temperature on respiration rates and does not account for a decrease in respiration observed at temperatures above 22 °C. Because of the Wisconsin Model equation's exponential temperature function, these problems cannot be fixed by changing parameter values. However, realistic calibrations of growth have been made with this formulation. The decrease in respiration by inactive fish at high temperatures observed by Myrick (1998) in laboratory respiration chambers may not be applicable in natural settings.

6.3.8. Other energy losses—

Many fish bioenergetic formulations include terms for energy losses owing to egestion, excretion, and specific dynamic action. InSTREAM does not include these terms because their effects are small compared to the uncertainties and variability in food availability and in the feeding and growth formulations (Bartell et al. 1986). These terms may be important at extremely low or high temperatures when the ability to digest food can limit growth; instead, InSTREAM uses the *cMax* function to limit food consumption at extreme temperatures.

6.3.9. Feeding strategy selection, net energy benefits, and growth—

The feeding strategy selection, net energy, and growth methods calculate a fish's daily growth for a specific habitat cell. Total food and energy intake is calculated and total energy losses subtracted, determining whether drift feeding or active searching is more profitable.

Variables with the word "food" in their name refer to prey (in g); "energy" variables refer to energy from prey (J). Prey energy density (the habitat parameter *habPreyEnergyDensity*, J/g) is used to convert grams of prey eaten to joules of energy intake. Hanson et al. (1997) provide values of *habPreyEnergyDensity* for various prey types. A value of 2500 J/g is reasonable for streams where drift prey is dominated by aquatic insect larvae; a value of 4000 J/g is appropriate for streams where drift is dominated by higher-energy prey such as amphipods. Parameter *habPreyEnergyDensity* applies to both drift and search food.

The energy density of fish (fish parameter *fishEnergyDensity*, J/g) is used to convert a fish's net energy intake to growth in weight. The energy density of salmonids actually varies through their life cycle (typically higher in adults, especially during gonad development prior to spawning), but this variation is ignored in InSTREAM. Literature summarized by Hanson et al. (1997) indicates that 5900 J/g is a reasonable value for all stream trout.

The following steps describe the process used by a fish to determine the feeding strategy it would use, and the resulting food intake and growth it would obtain, for a particular habitat cell. This process uses variables (e.g., *driftIntake*, *feedTime*, *searchIntake*) calculated using the methods described above.

1. Determine the daily drift intake that would be obtained in the absence of more dominant fish in the cell. This *dailyPotentialDriftFood* (g/d) is determined from the hourly intake rates and hours spent feeding:

$$dailyPotentialDriftFood \ = \ driftIntake \ \times \ feedTime.$$

2. Determine *dailyAvailableDriftFood*, the drift intake rate available after more dominant fish in the cell have consumed their intake.

3. Calculate the actual drift intake rate *dailyDriftFoodIntake* (g/d), considering whether it is limited by actual food availability or the physiological maximum intake, *cMax*:

$$dailyDriftFoodIntake \ = \ \min(dailyPotentialDriftFood,$$
$$dailyAvailableDriftFood, cMax).$$

4. Convert daily drift intake in grams of food to joules of energy, *dailyDriftEnergyIntake* (J/d):

$$dailyDriftEnergyIntake \ = \ dailyDriftFoodIntake \ \times \ habPreyEnergyDensity.$$

5. Conduct the bioenergetics energy balance to get net energy intake for drift feeding; total respiration (*respTotal*, J/d) depends on cell water velocity and whether the fish has access to velocity shelter:

$$dailyDriftNetEnergy \ = \ dailyDriftEnergyIntake \ - \ respTotal.$$

6. Determine the daily search-feeding intake that would be obtained in the absence of more dominant fish in the cell, *dailyPotentialSearchFood* (g/d):

$$dailyPotentialSearchFood \ = \ searchIntake \ \times \ feedTime.$$

7. Determine *dailyAvailableSearchFood*, the search intake available after more dominant fish have consumed their intake.

8. Calculate the actual search intake *dailySearchFoodIntake* (g/d), considering whether it is limited by food availability or maximum daily intake:

$$dailySearchFoodIntake \ = \ \min(dailyPotentialSearchFood,$$
$$dailyAvailableSearchFood, cMax).$$

9. Convert daily search intake to joules of energy, *dailySearchEnergyIntake* (J/d):

dailySearchEnergyIntake = *dailySearchFoodIntake* × *habPreyEnergyDensity*.

10. Conduct the bioenergetics energy balance to get net energy intake for search feeding:

dailySearchNetEnergy = *dailySearchEnergyIntake* − *respTotal*.

11. Select the most profitable feeding strategy by comparing *dailyDriftNetEnergy* to *dailySearchNetEnergy*; and determine the energy intake for the best strategy:

bestNetEnergy = max(*dailyDriftNetEnergy*, *dailySearchNetEnergy*).

12. Convert net energy intake to daily growth *dailyGrowth* (g/d):

dailyGrowth = *bestNetEnergy*/*fishEnergyDensity*.

13. Update the fish's weight:

fishWeight = *fishWeight* + *dailyGrowth*.

In the final step, *fishWeight* is set to zero if *dailyGrowth* is negative with a magnitude greater than *fishWeight* (this can happen in the model when growth is calculated for small fish in cells that would demand extremely high swimming speeds).

6.3.10. Preliminary parameter estimation for feeding and growth—
Many variables affect growth, so it must be calibrated incrementally. This section identifies ranges of values for food production that produce reasonable feeding and growth rates under simplified conditions. This preliminary parameter estimation makes it easier to calibrate growth in the whole model, where habitat selection and competition strongly affect growth.

This section discusses calibration of growth by adjusting the parameters for food production. The key food parameter, *habDriftConc*, can in fact be measured in the field instead of calibrated. However, we discourage attempting to use measured drift concentrations for several reasons. First, this parameter captures many of the uncertainties resulting from model simplifications such as ignoring variation in prey size and assuming fish feed only during daytime. Therefore, accurately measured drift concentrations may not produce accurate model results. Second, drift concentration measurements are expensive and uncertain; resources for field studies are probably better spent on fish data to calibrate the model.

Reasonable values of the search- and drift-food availability parameters are found by identifying ranges that meet criteria developed from field observations and laboratory growth data. These criteria are:

- Daily food intake under summer conditions should be in the range of 20 to 50 percent of *cMax*; *cMax* should rarely if ever limit food intake. This criterion is based on field research in which **average** food intake was estimated from observed growth and bioenergetics models. Railsback and Rose (1999), using a bioenergetics formulation similar to that used in InSTREAM, found food consumption by trout in relatively small streams of California's Sierra Nevada to average 30 to 35 percent of *cMax*. At these sites, temperatures were 15 to 19 °C; *cMax* increases sharply with temperature in this range. Individual fish in excellent habitat could have food intake well above the average. This criterion may not be valid in unusual situations where food is extremely abundant and trout growth rates very high, or at very low temperatures where *cMax* is very low.
- Drift feeding should be more profitable than active search feeding, except at low velocities, when turbidity is high, when benthic prey are extremely abundant, or for very small trout. Trout are rarely observed feeding only with the search strategy, and where both strategies are available, drift feeding is probably more often preferred (Nielsen 1992, Nislow et al. 1998).
- Growth under good conditions (high food intake, low swimming velocity) should not exceed growth rates observed in lab studies where fish were fed as much as they could eat (e.g., Myrick 1998, Myrick and Cech 2000). These lab growth rates are in the range of 2 to 6 percent of body weight per day, varying with temperature.

To estimate food parameter values, the entire feeding and growth formulation of InSTREAM was implemented in a spreadsheet. Parameter values from tables 9, 10, 11, and 12 were used. Summer conditions were represented: *feedTime* = 16 h and temperature = 15 °C. Turbidity was assumed to be zero. Both juvenile (5 cm length, 1.5 g weight) and adult (15 cm, 40 g) trout were simulated. Reasonable values for the drift-food parameter *habDriftConc* were identified as the range producing food intake (g/d) of 20 to 50 percent of *cMax* in the adult trout for trout using near-optimal velocities and velocity shelter. This range is 5×10^{-10} to 12×10^{-10} g/cm^3. Within this range of *habDriftConc*, modeled adult trout growth ranged 0.5 to 2.5 percent body weight per day. For 5-cm juvenile trout, this range of *habDriftConc* produced food intake between 50 and 100 percent of *cMax* and growth in the range of 5 to 15 percent per day; the lower ends of these ranges are consistent with observed rates.

The value of *habDriftRegenDist* was estimated by assuming a cell that contains 15-cm trout, each trout having a square territory 150 cm on each side (similar to Grant and Kramer 1990). The cell is also assumed to have a depth of 30 cm and velocity of 30 cm/s, approximating optimal feeding conditions. It was assumed that trout achieved food intake equal to 30 percent of *cMax*, or 0.11 g/h, and that under these conditions drift food production exactly equals consumption by the trout. With *habDriftConc* in the range of 5×10^{-10} to 7×10^{-10} g/cm^3, the value of *habDriftRegenDist* must be approximately 300 to 500 cm.

The assumptions used to estimate search intake parameters are (a) a search-feeding fish consumes the production of 2 m^2, so the value of *fishSearchArea* is 20,000 cm^2 and (b) a 5-cm trout can maintain growth of 0 to 2 percent body weight per day by search feeding for 16 h/d at 15 °C, at velocities of 10 cm/s or less. The range of *habSearchProd* values producing this growth range is 2×10^{-7} to 5×10^{-7} g $\cdot$ cm^{-2} $\cdot$ h^{-1}.

There are few published estimates of trout-food production rates with which to compare these parameter estimates. Published estimates of invertebrate production do not separate drift from invertebrates eaten at the benthic surface. The rate at which food drops in from overhead (included in food production in InSTREAM) is also rarely measured. Poff and Huryn (1998) reported overall food production rates in streams containing Atlantic salmon (*Salmo salar*) in the range of 4 to 24 g dry weight per m^2 per year, which converts to 10×10^{-7} to 60×10^{-7} g $\cdot$ cm^{-2} $\cdot$ h^{-1} (assuming a typical ratio of 20 for dry:wet weight; Hanson et al. 1997). The range of *habSearchProd* estimated above (2×10^{-7} to 5×10^{-7} g $\cdot$ cm^{-2} $\cdot$ h^{-1}) appears reasonable compared to this value: *habSearchProd* is expected to be a relatively small but not negligible fraction of the total production rate.

Examining how food intake and growth vary with cell water velocity helps to understand the feeding and growth formulation. Figure 13 illustrates how daily food intake (evaluated as the percentage of *cMax*) varies with velocity for both feeding strategies by 5-cm juveniles and 15-cm adult trout. Figure 14 illustrates the resulting growth (as percentage of body weight per day) and the effect of using velocity shelters on growth. These graphs assume a temperature of 15 °C, depth of 50 cm, feeding time is 16 h/d, *fishShelterSpeedFrac* is 0.3, *habDriftConc* is 5×10^{-10} g/cm^3, *habSearchProd* is 5×10^{-7} g $\cdot$ cm^{-2} $\cdot$ h^{-1}, *fishSearchArea* is 20,000 cm^2, *habPreyEnergyDensity* is 2500 J/g, and *fishEnergyDensity* is 5900 J/g. Figure 15 is identical to fig. 14 except for depicting winter conditions, with a temperature of 5 °C and feeding time of 12 h.

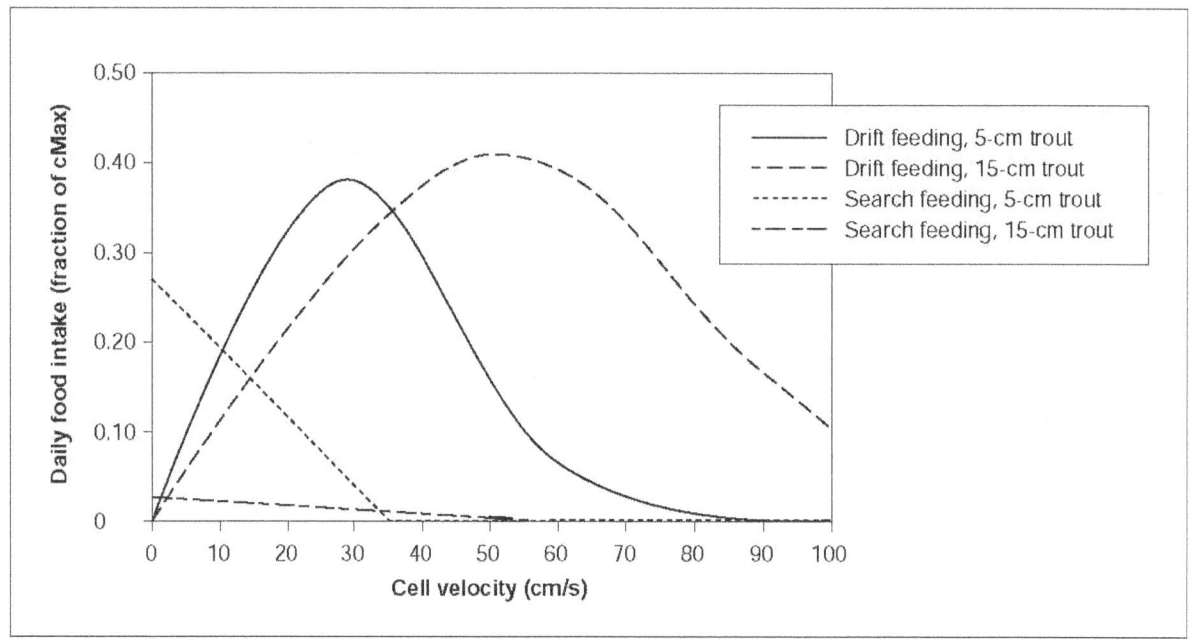

Figure 13—Variation in food intake with velocity for two sizes of trout, using drift or search feeding. Intake is depicted as percentage of *cMax* (physiological maximum daily intake).

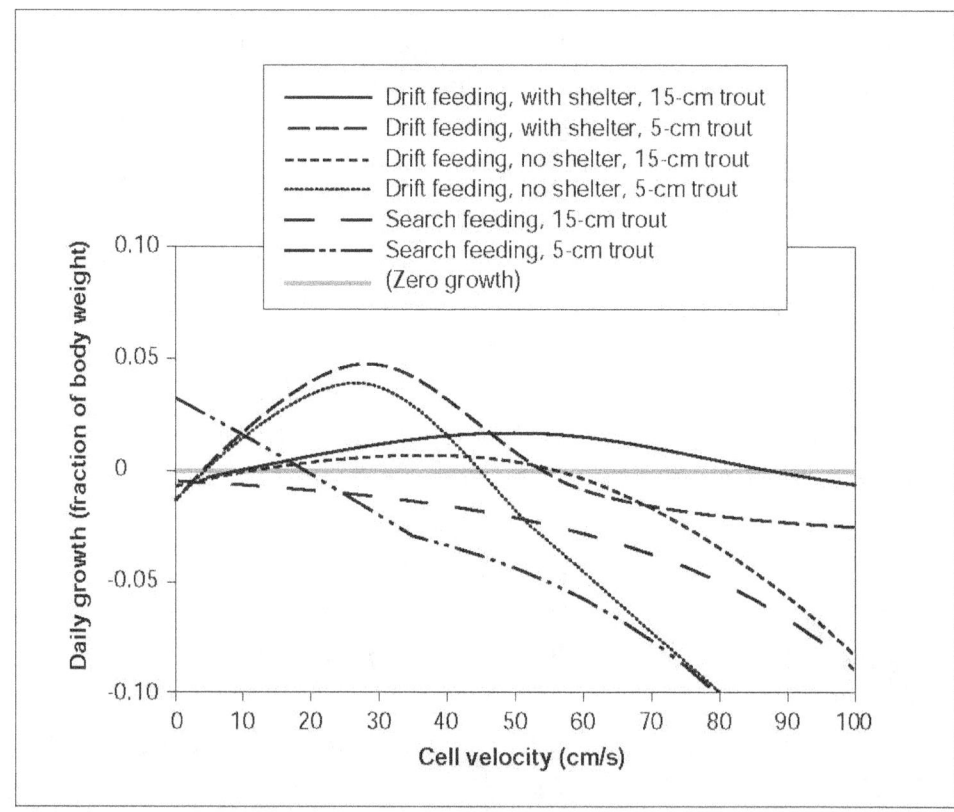

Figure 14—Variation in growth rate with velocity for two sizes of trout, drift and search feeding strategies, under summer conditions. Growth is depicted as percentage of body mass per day.

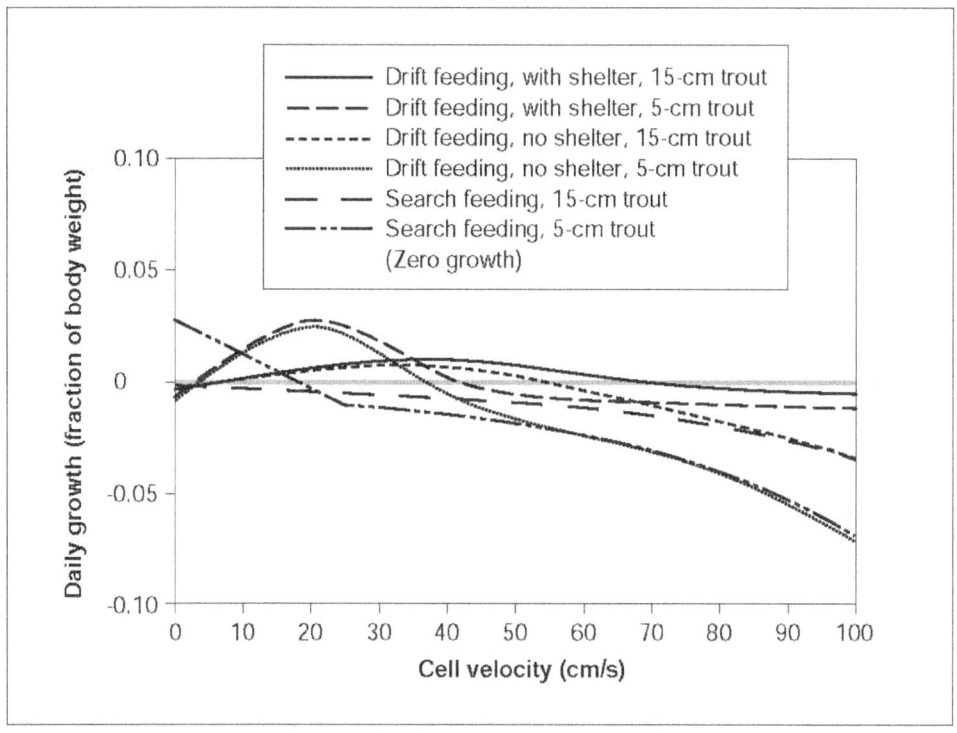

Figure 15—Variation in growth rate with velocity, under winter conditions.

Several patterns in these results are noteworthy in that they appear to reflect patterns observed in real trout:

- Conditions providing high intake do not always provide high growth, owing to the metabolic costs of swimming (especially for fish drift feeding without velocity shelters).
- The use of velocity shelters for drift feeding is very beneficial. Shelters increase the growth rate but also, more importantly, increase the range of velocities under which growth is positive.
- Larger fish can drift feed profitably over a wider range of velocities, and at higher velocities, than can smaller fish.
- Search feeding is a profitable strategy only for small fish in low velocities.
- The relative benefits of drift feeding increase with fish size.
- At low temperatures, growth is lower overall and is maximized at lower velocities.

6.4. Fish Survival

Survival simulations determine, each day, which fish die from what causes. The survival action for a fish is a two-step process. First is calculating the probability of surviving each of several mortality sources. Second is determining, stochastically, whether the fish actually dies from any of the mortality sources.

The survival methods simulate important **mortality sources**, which are represented in InSTREAM as **survival probabilities**: the daily probability of not being killed by one specific mortality source. InSTREAM includes these sources of mortality:

- High temperature
- High velocity (exhaustion and inability to maintain position)
- Stranding (including predation risk associated with extremely shallow habitat)
- Poor condition (starvation and disease when weight is low)
- Predation by terrestrial animals
- Predation by fish

The primary reason InSTREAM separately represents different mortality sources is that the probability of surviving each varies differently with fish state and habitat conditions. For example, the risk of predation by terrestrial animals is greatest for large fish in shallow, low-velocity cells; the risk of predation by fish is greatest for small fish in deep cells. The primary adaptive behavior represented in InSTREAM—habitat selection—depends on survival probabilities. For habitat selection to be modeled realistically, InSTREAM must represent how different mortality sources vary differently over time, among fish, and over space.

Survival probabilities are used for two purposes. First, survival probabilities strongly influence habitat selection (section 6.2). The second use, addressed here, is to model mortality: when and why each fish actually dies. The model uses the same methods to determine survival probabilities for both habitat selection and mortality.

Death of fish is modeled stochastically by comparing pseudo-random numbers to the survival probabilities. Potential death by each mortality source is treated as an independent event. On each simulated day, the model determines whether each fish dies of each mortality source using these steps:

- Calculate the survival probability from the current state of the fish and its cell.
- Obtain a pseudo-random number from a uniform distribution between 0 and 1.

- If the random number is greater than the survival probability, then the fish dies as a result of the mortality source. No further mortality sources are evaluated for the fish.
- If the fish does not die, evaluate the next mortality source.

Although the model treats death by each mortality source independently, the order in which mortality sources are evaluated can have a (usually very small) effect on how many fish die of each kind of mortality. We discuss the ordering of mortality sources with the model schedule in section 12.2.

It is important to understand that seemingly high daily survival probabilities can result in low survival over time. For example, a daily survival probability of 0.99 results in mortality of 26 percent of fish within 30 days ($0.99^{30} = 0.74$). Survival probabilities should be well above 0.99 if they are not to cause substantial mortality over time.

The following sections describe the detailed formulations used to calculate survival probabilities for each mortality source.

6.4.1. High temperature—

This mortality source represents the breakdown of physiological processes at high temperatures. It does not represent the effect of high temperatures on bioenergetics (reduced growth at high temperature). The high-temperature survival function is based on laboratory data collected from (presumably) disease-free fish, so it does not represent the effect of disease even though fish are probably more susceptible to disease at high temperatures. Instead, disease is modeled as part of poor condition mortality; a fish able to maintain its weight at sublethal temperatures is assumed to remain healthy.

Although input to InSTREAM includes only daily mean temperature, mortality is related to the daily maximum temperature as well as the mean (although the relative importance of mean v. maximum temperature is not clear: Dickerson and Vinyard 1999, Hokanson et al. 1977). The survival probability parameters therefore assume a difference between mean and peak temperatures. The temperature mortality parameters can be reevaluated for sites with particularly high or low diurnal temperature variations.

High-temperature mortality has been addressed by numerous laboratory studies, but models of this mortality remain variable and uncertain because mortality varies (a) with laboratory conditions and techniques and the endpoints used to define mortality, (b) between laboratory and field conditions, and (c) among individuals. Data from Behnke (1992) and Moyle and Marchetti (1992) indicate that any interspecific differences in lethal temperatures are not clearly distinguishable from uncertainty and variability in the measurements. Recent laboratory data show approximately 60-percent survival of golden trout (*Oncorhynchus mykiss*) juveniles over a 30-d period at a constant 24 °C (Myrick 1998), equivalent to a daily survival of 0.98. Dickerson and Vinyard (1999) measured survival of Lahontan cutthroat trout (*O. clarkii*) for 7 d at high temperatures, finding zero survival at 28 °C, 40 percent survival at 26 °C (equivalent to daily survival of 0.88), and 100 percent survival at 24 °C. These experiments indicate that high-temperature mortality can be modeled well as a logistic function. The parameters in table 13 (illustrated in fig. 16) appear suitable for sites with relatively low diurnal variation in temperature; they produce daily survival of 0.98 at 24 °C, 0.88 at 26 °C, and < 0.5 at 28 °C.

Table 13—Parameter values for high-temperature mortality

Parameter	Definition	Value
mortFishHiTT9	Daily mean temperature (°C) at which high-temperature survival is 90 percent	25.8
mortFishHiTT1	Daily mean temperature (°C) at which high-temperature survival is 10 percent	30.0

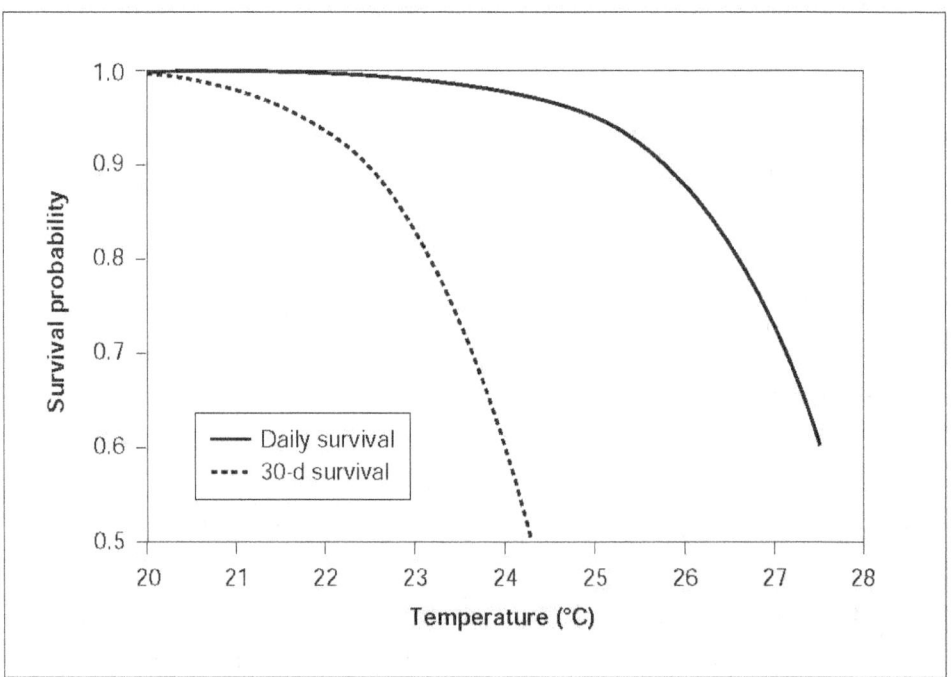

Figure 16—Survival probability function for high temperature. Survival is the probability of a trout surviving high temperature mortality for 1 day (solid line) and for 30 days (dashed line) (equal to the daily survival raised to the power 30).

6.4.2. High velocity—

The high-velocity survival function represents the potential for trout to suffer fatigue or lose their ability to hold position in a cell with high velocity. This function is included not because trout often die from high velocity, but because it strongly affects habitat selection: mortality from high velocities is not observed in nature because fish avoid it by moving. Velocities posing mortality risk can be widespread at high flows, but can also occur (especially for small fish) at normal flows.

The survival probability is based on the ratio of the swimming speed a fish uses in a cell to the fish's maximum sustainable swim speed. The swimming speed used in a cell is determined when calculating respiration energy costs (section 6.3.7): fish are assumed to swim at the cell's water velocity unless they are drift-feeding with access to velocity shelters. Fish using velocity shelters are assumed to swim at a speed equal to the cell's water velocity times the parameter *fishShelterSpeedFrac*.

Maximum sustainable swim speed (*maxSwimSpeed*, cm/s) is a particularly important state variable for model trout. As a component of both high-velocity mortality and drift feeding (section 6.3.3), *maxSwimSpeed* strongly affects the relationship between a cell's water velocity and habitat quality for various sized trout. Because InSTREAM uses a daily time step, the maximum swim speed used for high-velocity mortality must be a speed that fish can swim for hours, not a burst or short-term maximum speed. The formulation for *maxSwimSpeed* is based on literature values of "critical swimming speed" (often abbreviated as U_{crit}), a standard approach to estimating maximum sustainable speed in laboratory tests. Measurement of U_{crit} involves repeatedly stepping up the swimming speed and holding it for a specified time interval until the fish is exhausted; different time intervals can be used to estimate short-term versus long-term sustainable swim speeds. We used relatively long-term values of U_{crit} to model *maxSwimSpeed*. Trout may start to use white (fast-twitch) muscle fibers at 90 to 95 percent of U_{crit} (Myrick 1998). Therefore, 90 percent of the U_{crit} is a reasonable estimate of the speed fish can sustain for long periods (C. Myrick, Department of Fish, Wildlife, and Conservation Biology, Colorado State University, pers. comm. with S. Railsback, 10 May 1999).

U_{crit} for trout has been measured at different temperatures and fish lengths by a number of researchers. These studies examined brown (Butler et al. 1992), cutthroat (Hawkins and Quinn 1996, MacNutt et al. 2004), and rainbow and golden trout (Alsop and Wood 1997; Myrick and Cech 2000, 2003; Schneider and Connors 1982;

Taylor et al. 1996). [The study by Griffiths and Alderdice (1972) was not used even though it has been the basis of several previous models of maximum swimming speed. Griffiths and Alderdice measured juvenile coho salmon swimming speed over temperatures between 2 and 26 °C. However, they did not provide sufficient information to distinguish the effects of fish size and temperature and apparently did not control these two variables separately.]

Results of these studies differ considerably, likely because of differences in experimental equipment and techniques, and to variability in the exercise condition of the fish. However, two general conclusions can be drawn. First, *maxSwimSpeed* increases with fish length (fig. 17). Second, *maxSwimSpeed* varies nonlinearly with temperature, peaking at temperatures around 10 to 15 °C (fig. 18). The formulation for *maxSwimSpeed* therefore has two terms: the first represents how swimming speed at 10 to 15 °C varies with fish length, and the second modifies *maxSwimSpeed* for temperature.

$$maxSwimSpeed = [(fishMaxSwimParamA \times fishLength) \\ + fishMaxSwimParamB] \times [(fishMaxSwimParamC \times temperature^2) \\ + (fishMaxSwimParamD \times temperature) + fishMaxSwimParamE].$$

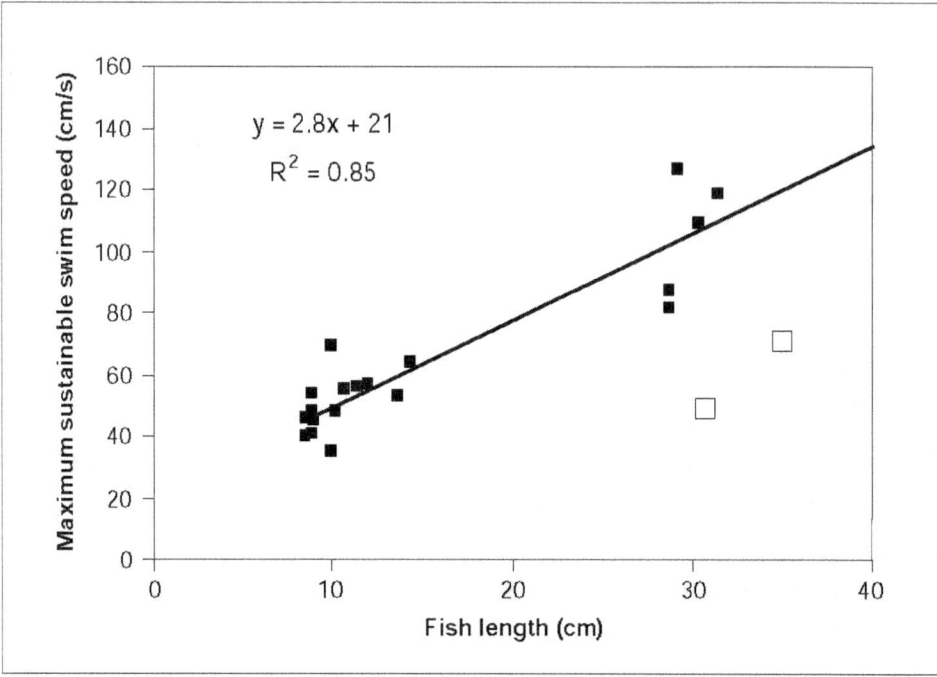

Figure 17—Maximum sustainable swimming speed as a function of fish length; measurements made at 10 to 15 °C. The points marked as open squares were omitted as outliers.

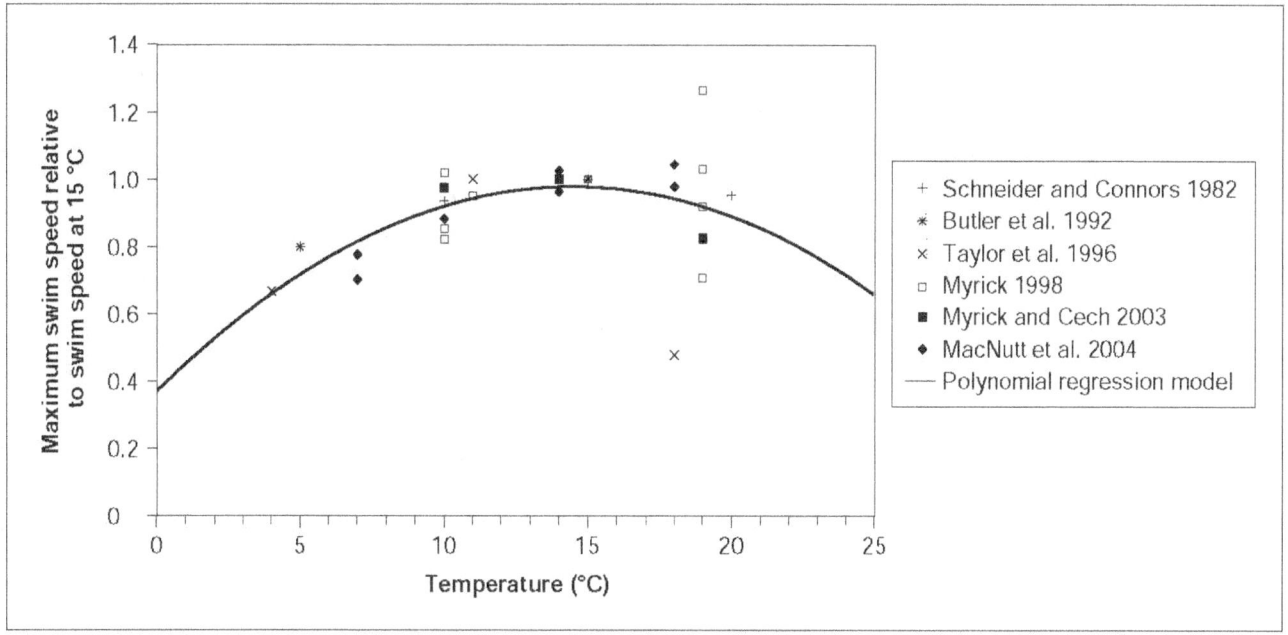

Figure 18—Variation in maximum sustainable swim speed with temperature. Observations from four studies are shown separately. The Y axis is the measured swim speed divided by the speed measured at (or near) 15 °C in the same study.

Table 14 provides parameter values fit to data from the studies cited above. Observations of U_{crit} from these studies were converted to maximum sustainable swimming speeds by multiplying U_{crit} by 0.9. The relation between *maxSwimSpeed* and trout length (parameters *fishMaxSwimParamA* and *fishMaxSwimParamB*) was fit using observations made at temperatures between 10 and 15 °C (fig. 17). A few of these literature values were omitted as outliers (as shown in the figures) because

Table 14—Parameter values for high-velocity mortality

Parameter	Definition	Value
fishMaxSwimParamA	Length coefficient in maximum swim speed equation (1/s)	2.8
fishMaxSwimParamB	Constant in maximum swim speed length term (cm/s)	21.0
fishMaxSwimParamC	Temperature squared coefficient in maximum swim speed equation ($°C^{-2}$)	-0.0029
fishMaxSwimParamD	Temperature coefficient in maximum swim speed equation ($°C^{-1}$)	0.084
fishMaxSwimParamE	Constant in maximum swim speed temperature term (unitless)	0.37
mortFishVelocityV9	Ratio of fish swimming speed to maximum swim speed at which high-velocity survival is 90 percent (unitless)	1.4
mortFishVelocityV1	Ratio of fish swimming speed to maximum swim speed at which high-velocity survival is 10 percent (unitless)	1.8

they appeared to underestimate swim speed. Parameters *fishMaxSwimParamC*, *fishMaxSwimParamD*, and *fishMaxSwimParamE* were fit via polynomial regression of (a) the ratio of swim speed at a temperature to swim speed at 15 °C in the same study, versus (b) temperature (fig. 18).

A decreasing logistic function relates survival probability to the fish's swimming speed in its habitat cell divided by the fish's *maxSwimSpeed* (fig. 19). The parameters for this function (table 14) are chosen so that high-velocity mortality is negligible at swimming speeds less than *maxSwimSpeed*, reflecting that (a) the laboratory equipment for measuring swim speeds does not provide the kinds of turbulence and fine-scale velocity breaks that trout can often use to reduce swimming effort in natural conditions and (b) stream fish are likely to be in better condition than laboratory fish.

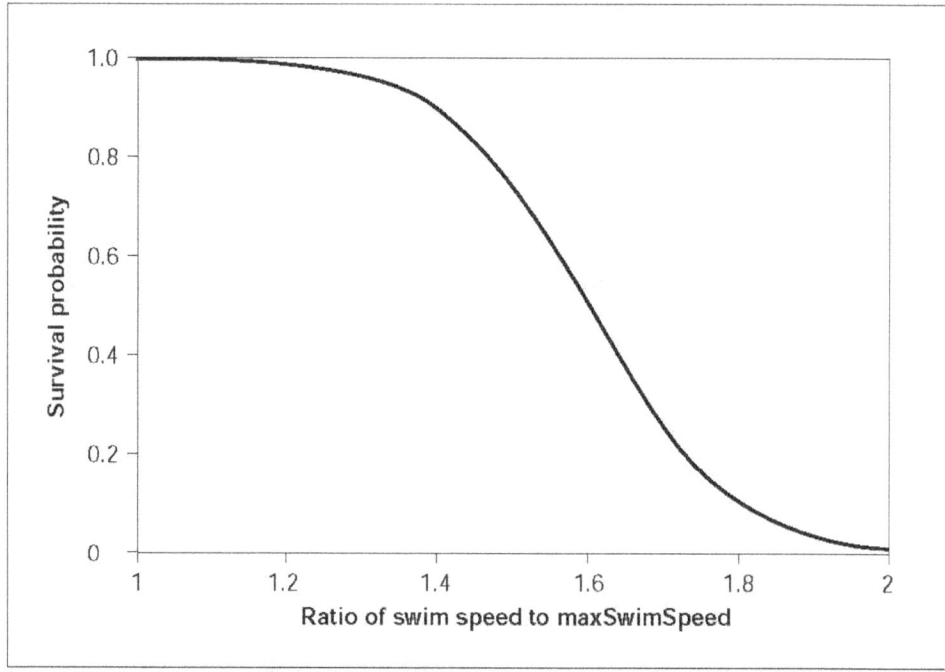

Figure 19—Survival probability function for high velocity. The X axis is the fish's actual swimming speed divided by its maximum sustainable swimming speed.

6.4.3. Stranding—

Stranding mortality represents the death of fish that are unable to move out of cells that become extremely shallow or dry as flow decreases. Fish in InSTREAM already have strong incentives to avoid cells with near-zero depth: low drift food intake and high risk of terrestrial predation. However, there can be cases where (a) a fish is prevented from reaching a cell with non-zero depth by its maximum movement distance or (b) no better habitat is available for other reasons.

Survival of stranding is modeled as an increasing logistic function of depth divided by fish length (fig. 20, table 15). Because the terrestrial predation function does not represent the greatly increased likelihood of predation when depth is extremely low (e.g., when fish are trapped in isolated pools (Harvey and Stewart 1991)), this risk is included as part of stranding mortality. The stranding survival function does not distinguish whether fish in very low or zero depths die from lack of water or from predation.

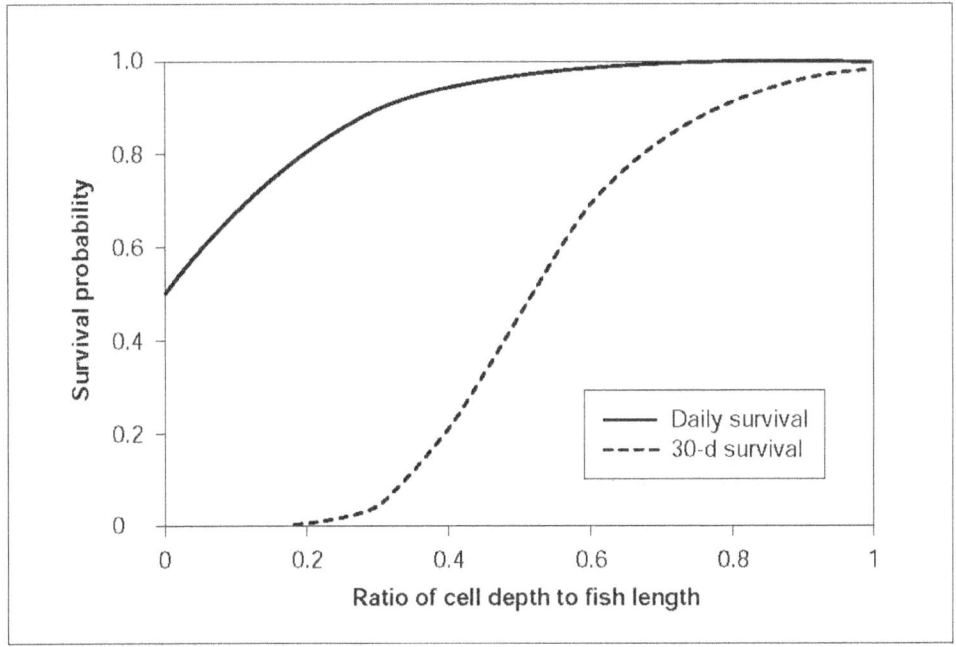

Figure 20—Survival probability function for stranding, showing the probability for surviving 1 day (solid line) and for 30 days (dashed line).

Table 15—Parameter values for stranding mortality

Parameter	Definition	Value
mortFishStrandD1	Ratio of depth to fish length at which stranding survival is 10 percent (unitless)	-0.3
mortFishStrandD9	Ratio of depth to fish length at which stranding survival is 90 percent (unitless)	0.3

The stranding parameters do not cause survival to reach zero when depth is zero, reflecting that real habitat (as opposed to the model's cells) has variation in bottom elevation. Some water could remain even if a cell's simulated depth becomes zero. Depth is divided by fish length to scale how the risks of low depths vary with fish size; shallow habitat that may be very valuable for small fish (protecting them from aquatic predation) can pose a stranding risk for large fish.

6.4.4. Poor condition—

Fish in poor condition (low value of the condition factor K, section 6.3.1) suffer greater risk of starvation, disease, and predation. These risks are combined in the poor condition survival probability. Simpkins et al. (2003a, 2003b) studied starvation mortality in large juvenile trout, finding:

- Trout can survive for long periods (over 147 d, in some cases) with no food intake.

- Survival is lower at higher swimming activity and temperature (which both increase metabolism).

- Relative weight (equivalent to K) decreased linearly over time during starvation.

- Mortality was predicted better by an index of lipid content than by K, in part because water replaces lipids as energy stores are depleted.

Unfortunately, modeling depletion of body lipids and related processes would add considerable complexity and uncertainty to InSTREAM, as they are not well understood. Instead, poor condition survival probability is represented as an increasing logistic function of K with parameter values estimated to provide reasonable survival probabilities over several days and weeks (fig. 21, table 16). The parameters produce a survival probability less than 100 percent even when K is at its maximum of 1.0, because disease can occur with low probability even when condition is relatively good.

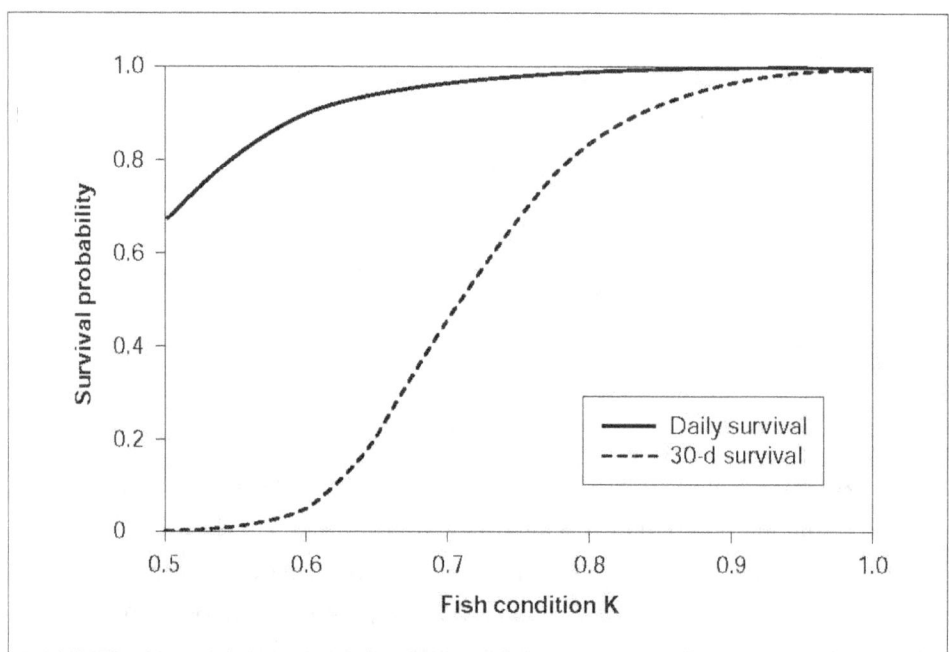

Figure 21—Survival probability function for poor condition. The dotted line is the probability for surviving for 30 days at a condition valued at K.

Table 16—Parameter values for poor condition mortality

Parameter	Definition	Value
mortFishConditionK1	Fish condition factor K at which survival is 10 percent (unitless)	0.3
mortFishConditionK9	K at which survival is 90 percent (unitless)	0.6

Poor condition is a unique mortality source in that fish cannot increase their survival probability immediately by selecting different habitat. Fish in poor condition have a strong incentive to select habitat that provides growth so their condition improves. However, sufficient growth to recover high condition takes a number of days. Even apparently high daily survival probabilities for this mortality source (e.g., 0.90) result in low probability of surviving until normal weight can be regained. As figure 21 indicates, the probability of surviving for extended periods becomes quite low when *K* falls below 0.8.

Before modifying the parameters for poor condition, users of InSTREAM should be aware that poor condition mortality can have a strong effect on habitat selection (section 6.2.2) as well as mortality. As a consequence, changes in parameter values are likely to have widespread, complex, and unexpected effects. For example, one might assume that increasing the survival probability (e.g., by decreasing *mortFishConditionK9* from 0.7 to 0.6) would result in less mortality owing to poor condition. However, because fish select habitat using a tradeoff between poor condition and other mortality sources (primarily predation), this change in parameters could result in fish selecting different habitat that has lower growth and lower predation risk, at least partially offsetting the expected reduction in poor condition mortality.

6.4.5. Terrestrial predation—

Predation by terrestrial animals is a dominant source of mortality to trout, especially adults (Alexander 1979, Harvey and Stewart 1991, Metcalfe et al. 1999, Quinn and Buck 2001, Valdimarsson et al. 1997). The terrestrial predation formulation represents predation by a mix of such predators as otters, raccoons, snakes, herons, mergansers, kingfishers, and dippers. Common characteristics of terrestrial predators that affect the survival probability function include that they are:

- Bigger than trout.
- Poorer swimmers than adult trout.
- Warm-blooded.
- Capable of locating fish prey from above the water's surface.

These characteristics differ among predators, but lead to these generalizations about terrestrial predation:

- Big trout are often more vulnerable than small trout.
- Risks extend year-round because warm-blooded predators may feed as much or more in winter (except those that hibernate or migrate).
- Risk increases when fish are more visible from above.

The formulation assumes a minimum survival probability that applies when fish are most vulnerable to terrestrial predation, and a number of "survival-increase functions" that can increase the probability of survival above this minimum. Survival-increase functions have values between 0 and 1, with higher values for greater protection from predation. The survival-increase functions are assumed to act independently. Therefore, terrestrial predation survival probability (*terrPredSurv*) is obtained by increasing the minimum survival (decreasing the difference between minimum survival and 1.0) by the **maximum** of the independent survival increase functions. This assumption is expressed mathematically as:

$$terrPredSurv = mortFishTerrPredMin + [(1 - mortFishTerrPredMin) \times \max(terrPredDepthF, terrPredTurbidityF, terrPredLengthF ...)].$$

where *terrPredDepthF*, *terrPredTurbidityF*, etc. are the values of the survival increase functions described below.

Under this approach, the value of *terrPredSurv* does not vary with the number of survival increase functions; only one function influences survival, the one providing the maximum survival increase. Survival increase functions can be added, removed, or revised without recalibrating the overall predation survival rate. However, the approach does not represent the potential combined effects of, for example, using deeper and faster habitat. Both depth and velocity make fish more difficult to see, and the combination of deep and fast is safer than only deep or fast, but this formulation does not represent the combined effect.

The value of *mortFishTerrPredMin* is assumed to be the daily probability of surviving terrestrial predation under conditions where the survival increase functions offer no reduction in risk. Field data for estimating this minimum survival are unlikely to be available, so it is best estimated by calibrating the model to observed abundance and habitat use patterns.

The following survival increase functions are included. (The effect of any function can be turned off by setting its function's parameters to yield values near zero.) We provide suggested parameter values at the end of the section (table 17).

Table 17—Parameter values for terrestrial predation mortality

Parameter	Definition	Value
mortFishTerrPredMin	Daily survival probability owing to terrestrial predators under most vulnerable conditions (unitless)	0.99 (until fit via calibration)
mortFishTerrPredD1	Depth at which survival increase function is 10 percent of maximum (cm)	Small streams: 5 Large rivers: 50
mortFishTerrPredD9	Depth at which survival increase function is 90 percent of maximum (cm)	Small streams: 150 Large rivers: 300
mortFishTerrPredL9	Fish length at which survival increase function is 90 percent of maximum (cm)	3
mortFishTerrPredL1	Fish length at which survival increase function is 10 percent of maximum (cm)	6
mortFishTerrPredF9	Feeding time at which survival increase function is 90 percent of maximum (h)	0
mortFishTerrPredF1	Feeding time at which survival increase function is 10 percent of maximum (h)	18
mortFishTerrPredV1	Velocity at which survival increase function is 10 percent of maximum (cm/s)	Small streams: 20 Large rivers: 20
mortFishTerrPredV9	Velocity at which survival increase function is 90 percent of maximum (cm/s)	Small streams: 100 Large rivers: 300
mortFishTerrPredH9	Distance to hiding cover at which survival increase function is 90 percent of maximum (cm)	-100
mortFishTerrPredH1	Distance to hiding cover at which survival increase function is 10 percent of maximum (cm)	500
mortFishTerrPredT1	Turbidity at which survival increase function is 10 percent of maximum	10
mortFishTerrPredT9	Turbidity at which survival increase function is 90 percent of maximum	50

First, however, **an important note on parameter sensitivity**: the sensitivity analyses discussed in chapter 3 show that InSTREAM results can be quite sensitive to the parameters that define how terrestrial predation risk depends on habitat variables. This sensitivity is not surprising, considering that terrestrial predation is normally the only mortality source for adult trout. Particularly important are the parameters that define the survival increase functions for habitat parameters, especially *mortFishTerrPredD9, mortFishTerrPredV9, mortFishTerrPredH9,* and *mortFishTerrPredH1.* If these parameters are set in such a way that the survival increase function is very close to 1.0 in several or many cells, then trout occupying those cells can be almost immune to mortality. For example, if the "small stream" parameters for depth illustrated in figure 22 were used in a large river with many cells having depth greater than 200 cm, then trout in these cells would have very low terrestrial predation risk and could live for many years. Changing the parameter *mortFishTerrPredD9* could greatly change the amount of habitat where predation risk is very low. (In reality, rivers with extensive deep water also likely have predators that can be effective in deep water.) This issue also applies if velocity and distance to hiding cover survival increase functions are very steep and near 1.0 for some cells; some parts of the simulated habitat can be nearly risk-free, producing higher populations of adult trout.

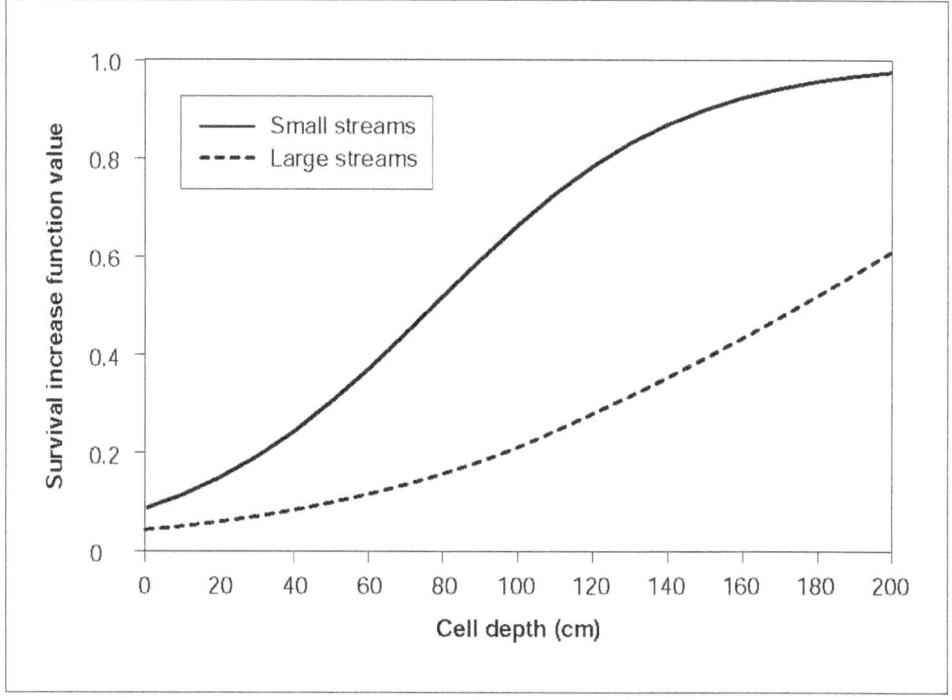

Figure 22—Depth survival increase function for terrestrial predation survival.

Depth—Fish are more vulnerable to terrestrial predators when in shallow water, where they are easier for predators to locate and catch (Harvey and Stewart 1991). The depth survival increase function is an increasing logistic curve: survival increases as depth increases (fig. 22). Power (1987) indicated that predation by birds is low at depths greater than 20 cm, and Hodgens et al. (2004) reported that 85 percent of successful strikes by herons occurred at depths less than 20 cm, although some were at depths up to 50 cm. However, predators that are larger or better swimmers (mergansers, otters) are effective at greater depths, especially in clear water. (Note that the very high risk of terrestrial predation that occurs when fish are in near-zero depths is included in stranding mortality.)

Appropriate values for the depth survival increase function parameters can differ among sites. Parameters useful in relatively small streams of coastal California (Railsback and Harvey 2001) provide high relative survival in depths > 1 m. However, these parameters were not useful for the much larger Green River in Utah, for example, where depths can be several meters and otters are prevalent. Figure 22 illustrates parameter values for small streams and large rivers (table 17).

Turbidity—Because turbidity makes fish less visible to terrestrial predators, it is assumed to be an important survival increase function. In the absence of quantitative data, this formulation considers the observed effect of turbidity on the ability of fish to detect prey (section 6.3.3), which shows the ability to detect drifting invertebrates declining toward zero at 40 NTU. Fish are likely more visible than invertebrates because of their size, but terrestrial predators must observe prey through greater distances of water than must fish predators. Therefore, the turbidity survival increase function has little effect at values below 5 NTUs but strongly reduces terrestrial predation risk at >40 NTU (fig. 23).

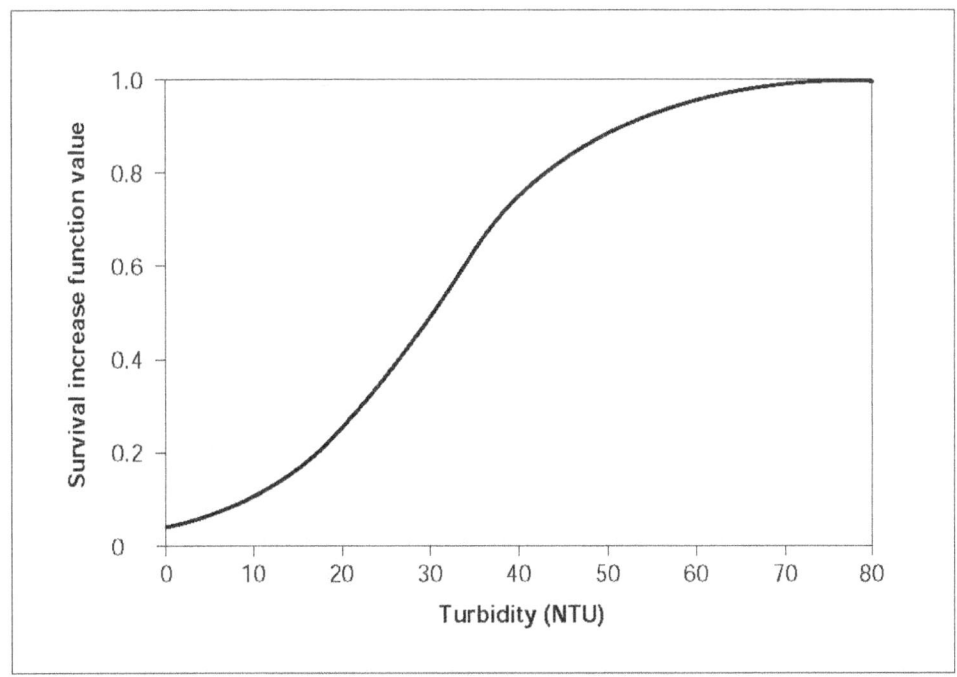

Figure 23—Turbidity survival function for terrestrial predation survival.

Fish length—Small fish are less vulnerable to terrestrial predation (Harvey and Stewart 1991), presumably because they are less visible (Power 1987), less desirable, and possibly more difficult to capture, than larger fish. However, dippers (*Cinclus mexicanus*) select trout fry and other small fish (Thut 1970), so very small fish are not invulnerable to terrestrial predation. Therefore, survival of terrestrial predation is assumed to decrease with fish length, but only fish less than 4 cm in length are relatively protected (fig. 24). These parameter values should be reconsidered for sites where predation is dominated by larger mammals (otters, bears) that strongly prefer large fish.

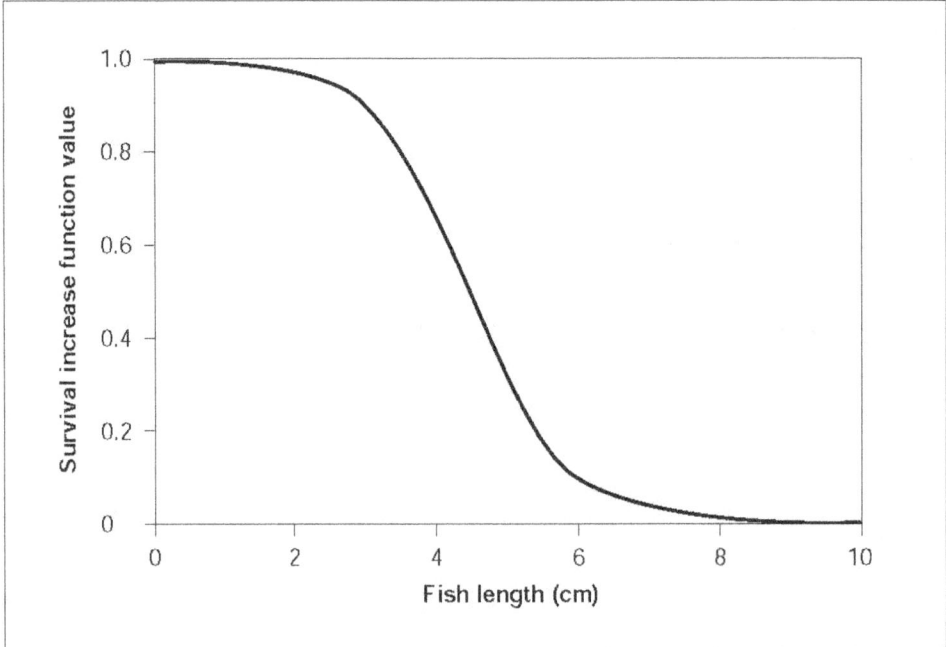

Figure 24—Fish length survival increase function for terrestrial predation survival.

Feeding time—Fish are much more vulnerable to predation when they actively feed during the day versus when they are resting and hiding at night (Metcalfe et al. 1999). The survival increase function is modeled as a decreasing function of *feedTime* (h), the hours spent feeding per day (section 6.3.2). Parameters are chosen so survival decreases nearly linearly with *feedTime* (fig. 25).

Water velocity—Water velocity is assumed capable of increasing terrestrial predation survival because (1) velocity-caused turbulence makes fish harder to see and (2) some predators are poorer swimmers than trout so they are expected to be less able to capture fish in faster water. The survival increase function is therefore an

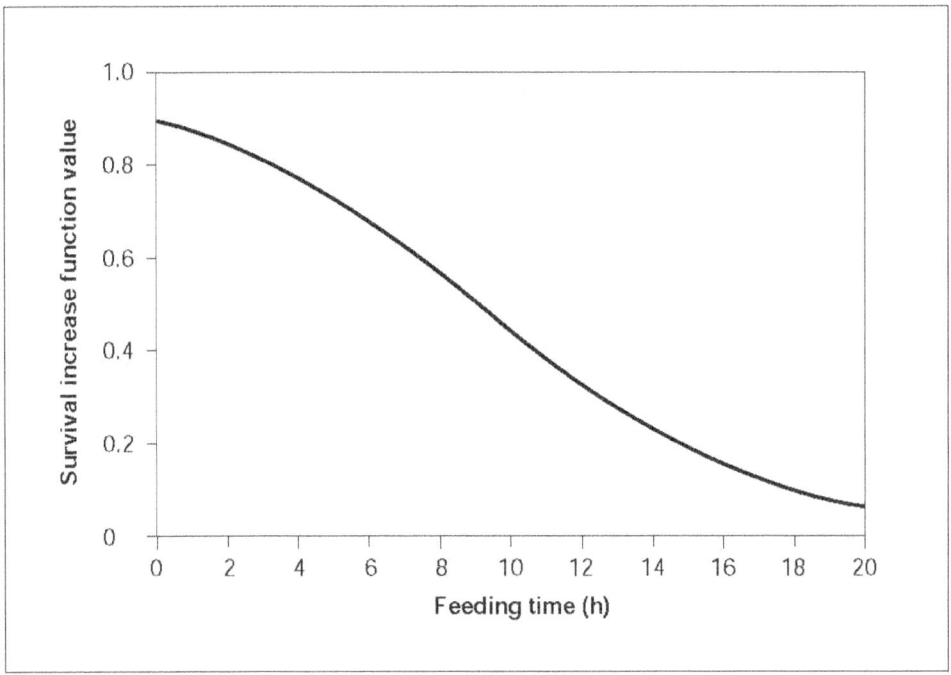

Figure 25—Feeding time function for terrestrial predation mortality.

increasing logistic curve that provides sharply increasing protection from terrestrial predators at velocities above 50 cm/s (fig. 26). As with the depth survival increase function, useful parameter values for the velocity function may differ between small and large streams. In small streams, high velocities combine with high turbulence and obstacles to make swimming difficult. In large rivers, however, habitat with high velocity and low turbulence may be common, so good swimmers such as mergansers and otters may perform quite well. Two sets of parameter values are provided in table 17 and illustrated in figure 26.

Temperature—No temperature-based survival increase function is included in InSTREAM because there are no clear mechanisms that would cause terrestrial predation pressure (unlike fish predation) to change with temperature. There is not a good basis for assuming predator activity is lower in winter; most important terrestrial predators are warm-blooded and many do not hibernate. In fact, such predators need additional food to maintain their metabolic needs in winter. The reduced swimming ability of trout at low temperatures can also offset any decrease in predator activity or density by reducing the ability of trout to escape (Metcalfe et al. 1999). Terrestrial predation can be greatly reduced when rivers freeze over, but ice is not represented in InSTREAM.

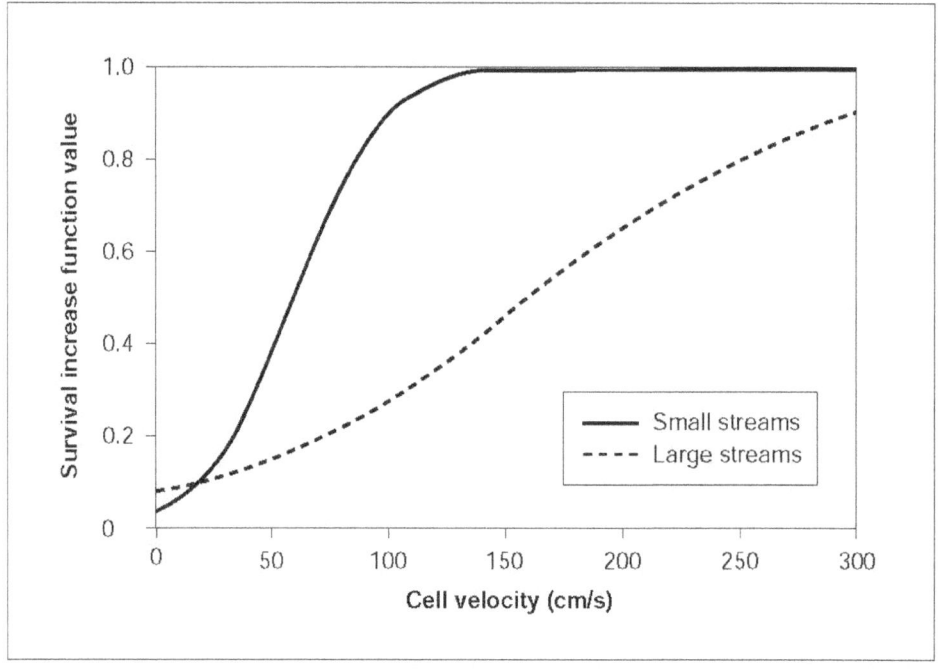

Figure 26—Velocity survival increase function for terrestrial predation survival, with parameters for both small streams and large rivers.

Distance to hiding cover—Fish can avoid mortality by hiding when predators are detected. The success of this tactic depends on the presence of hiding cover and the distance fish must travel to reach it. The value of hiding cover clearly occurs at a spatial scale different from the cell size typically used in InSTREAM; hiding cover several to tens of meters from a fish can provide at least some protection from predation.

Hiding cover is represented with a survival increase function that increases as distance to hiding cover decreases. Distance to cover (*cellDistToHide*, cm) is an input for each habitat cell, estimated in the field as the average distance a fish in the cell would need to move to hide from a predator. The value of *cellDistToHide* can range from near zero, for cells where a bottom of boulders or vegetation provides almost continuous cover, to many meters for cells lacking bottom cover. Very short distances to hiding cover (< 100 cm) provide nearly complete protection from some predators, but do not protect fish from predators that strike quickly (e.g., some birds) or that could be able to extract trout from hiding (e.g., otters). Cover several meters away is still valuable for escaping from terrestrial predators that have been detected. Therefore, the effect of distance to hiding cover is modeled as a decreasing logistic function of *cellDistToHide* (fig. 27).

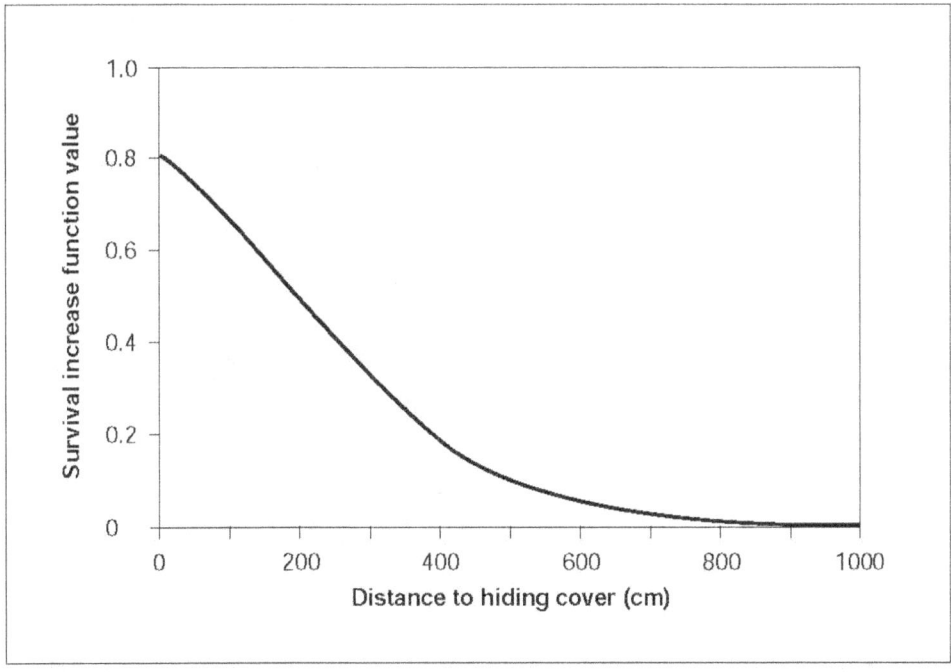

Figure 27—Distance to hiding cover function for terrestrial predation survival.

6.4.6. Aquatic predation—

The aquatic predation formulation represents mortality owing to predation by fish. In some trout populations, aquatic predation is only or primarily via cannibalism by large trout; in other populations predation may be from species not otherwise represented in the model (e.g., bass [*Micropterus* spp.] or pikeminnow [*Ptychocheilus* spp.]). The example parameter values given here were chosen for trout-only sites, but aquatic predation parameters can be adjusted to represent the presence of nontrout species. Such parameter changes should be based on knowledge of the predators' physiology and behavior. For example, a predator such as bass may have a larger gape and be less active at low temperature than trout. Hence, aquatic predation parameters could be adjusted so trout density has little effect, larger trout are more vulnerable, and risk drops more rapidly as temperature increases. It is unlikely that these parameter values can be estimated via calibration of inSTREAM's mortality results to empirical data because accurate data on rates and causes of mortality, especially for juvenile trout, is unlikely to be available.

The formulation can represent the effect of adult trout density on aquatic predation survival, making this survival probability the only component of InSTREAM with **direct** density dependence.

As with terrestrial predation, the formulation uses a minimum survival probability that applies when fish are most vulnerable to aquatic predation, and a number of survival increase functions:

$$aqPredSurv = mortFishAqPredMin + [(1 - mortFishAqPredMin) \times \max(aqPredDepthF, aqPredLengthF, aqPredVelF ...)]$$

where *aqPredSurv* is the daily survival probability for a particular fish in a particular habitat cell and *aqPredDepthF*, *aqPredLengthF*, etc. are survival increase function values. The value of *mortFishAqPredMin* is the daily probability of surviving aquatic predation under conditions where the survival increase functions offer no reduction in risk. As with terrestrial predation, data for directly estimating aquatic risks are unlikely to be available, so it is recommended that *mortFishAqPredMin* be estimated by calibrating the model to observed patterns of abundance and habitat selection by juvenile fish.

Especially at sites where trout rarely get larger than 20 to 30 cm, cannibalism by trout appears to be rare. At Little Jones Creek, less than 1 percent of adult fish contained juveniles (Railsback and Harvey 2001). However, the risk of predation appears to be an important factor driving habitat selection: fish commonly respond to larger piscivorous fish by shifting to shallow water (e.g., Brown and Moyle 1991). If aquatic predation rarely occurs, it is likely because small fish avoid it by avoiding risky habitat. Also, there have been anecdotal reports of very high cannibalism rates during fry emergence in some salmonids. A value of 0.9 can be used as a precalibration estimate of *mortFishAqPredMin*.

There is no survival increase function for distance to hiding cover in the aquatic predation formulation. This decision was made because only small trout are usually vulnerable to aquatic predators, and small trout are capable of hiding in many places that do not offer refuge to adult trout (e.g., between relatively small cobbles) and are not included in the estimation of distance to hiding cover.

The aquatic predation survival formulation includes the following survival increase functions. Table 18 provides potential parameter values.

***Predator density*—**This function represents how survival of aquatic predation depends on the density of **trout** predators. (It is important to understand that this function represents only the effect of trout included in the model and not other piscivorous fishes. Parameters can be chosen to minimize this function if nontrout fish dominate aquatic predation.) When adult trout abundance is greatly reduced, juveniles can safely use a wider range of habitat and, hence, have greater growth and survival to adulthood. The predator density survival increase function causes

Table 18—Parameter values for aquatic predation mortality at sites where adult trout dominate fish piscivory

Parameter	Definition	Value
mortFishAqPredMin	Daily survival probability owing to aquatic predators under most vulnerable conditions (unitless)	0.9 (until fit via calibration)
fishPiscivoryLength	The length at which trout become capable of preying on other trout (cm)	15
mortFishAqPredP9	Predator density at which survival increase function is 90 percent of maximum (cm^{-2})	2×10^{-6}
mortFishAqPredP1	Predator density at which survival increase function is 10 percent of maximum (cm^{-2})	1×10^{-5}
mortFishAqPredD9	Depth at which survival increase function is 90 percent of maximum (cm)	5
mortFishAqPredD1	Depth at which survival increase function is 10 percent of maximum (cm)	20
mortFishAqPredL1	Fish length at which survival increase function is 10 percent of maximum (cm)	4
mortFishAqPredL9	Fish length at which survival increase function is 90 percent of maximum (cm)	8
mortFishAqPredF9	Feeding time at which survival increase function is 90 percent of maximum (h)	0
mortFishAqPredF1	Feeding time at which survival increase function is 10 percent of maximum (h)	18
mortFishAqPredT9	Temperature at which survival increase function is 90 percent of maximum (°C)	2
mortFishAqPredT1	Temperature at which survival increase function is 10 percent of maximum (°C)	6
mortFishAqPredU9	Turbidity at which survival increase function is 90 percent of maximum (NTU)	80
mortFishAqPredU1	Turbidity at which survival increase function is 10 percent of maximum (NTU)	5

NTU = nephelometric turbidity units.

the survival increase function to increase as the density of piscivorous trout decreases. Two additional assumptions are needed to implement this function.

First, piscivorous trout must be defined. Any trout with length greater than the parameter *fishPiscivoryLength* (cm) is assumed to be a potential predator on smaller trout (and referred to here as a "piscivorous trout"). This is a simplification, because in reality the larger a fish becomes, the larger its potential prey. Considering observed predator-prey size ratios for salmonids (Keeley and Grant 2001), values in the range of 15 to 30 cm are reasonable for *fishPiscivoryLength*. This parameter should be considered site-specific; trout may be piscivorous only at larger sizes in fertile streams where other prey are abundant (Keeley and Grant 2001).

The second additional assumption concerns the spatial scale over which trout predation is represented. Predator density could be represented in InSTREAM at the cell, reach, or multiple-reach scales. We chose the reach scale because large, piscivorous trout are likely to forage among habitat cells. Therefore, predator density in this survival increase function is defined as the number of trout in the reach with length greater than *fishPiscivoryLength*, divided by the area (cm^2) of the reach (section 5.1.1). This density is evaluated using the density of piscivorous trout during the current time step; it is updated after all piscivorous trout have selected their habitat.

Parameters for the logistic decrease in survival with increasing predator density depend on whether the modeled trout are the only piscivorous fish. The parameters illustrated in figure 28 represent a site lacking other piscivores. The parameters reflect (a) near-zero risk in the absence of piscivorous trout and (b) a steep decline in survival as predator density exceeds 1 piscivorous trout per 25 m^2 (250,000 cm^2) of reach area. Post et al. (1998) measured the mortality of tethered juvenile trout owing to predation by adult trout in lakes. Risk increased exponentially with adult trout density, rising very sharply between 8 and 10 predators per 1000 m^3. This result supports a logistic-like relation between adult trout density and juvenile trout survival probability, but the exact relation is not directly applicable to InSTREAM because (a) it was obtained in lakes where cover and other habitat complexities may mediate the effect of predator density and (b) risks were evaluated over 1-hour periods, whereas InSTREAM model uses a daily time step.

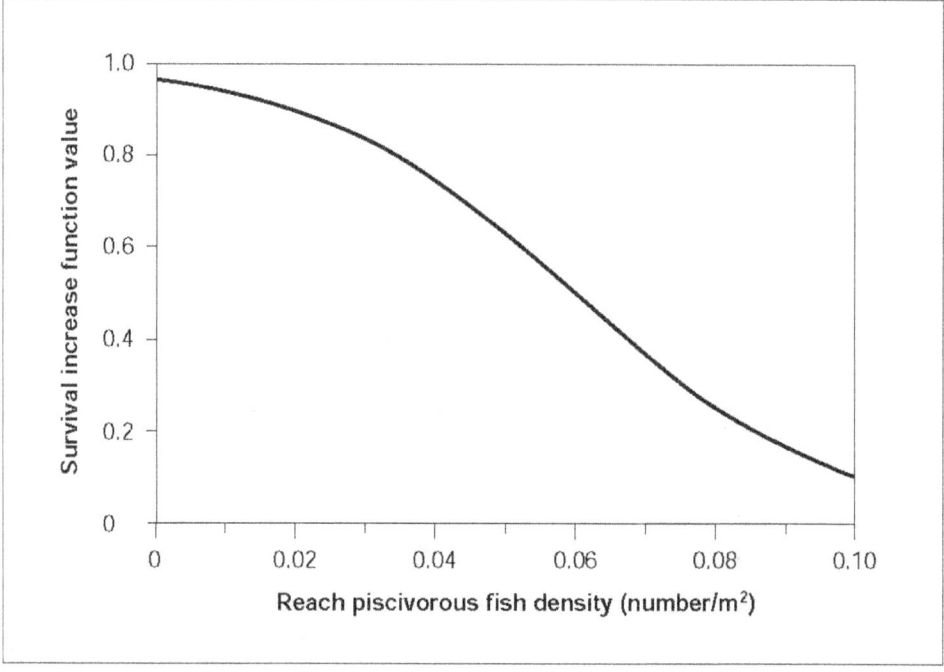

Figure 28—Predator density survival increase function for aquatic predation survival. The X axis is the density (number per m^2) of piscivorous trout in the fish's reach.

For sites where fish other than the trout represented in the model pose a piscivory risk, parameter values should be adjusted to reflect the reduced importance of trout to survival of aquatic predation. For example, if a site has a dense population of piscivorous pikeminnow, then trout density may have little effect on survival. In that case, the predator density function should be low and relatively flat (e.g., *mortFishAqPredP9* = -1.0; *mortFishAqPredP1* = 0.001).

Finally, for some trout there is a subtle and typically negligible difference between the aquatic predation survival probability they use in their habitat selection decisions (section 6.2.3) and the probability they are actually exposed to during mortality simulations. During habitat selection, the simulated trout (from largest to smallest) decide which habitat reach and cell to occupy, considering (along with growth) all potential mortality sources. For a piscivorous trout making its habitat selection decision, it is impossible to know what the aquatic predation risk is exactly because the density of piscivorous trout in the reach is not yet known: smaller trout that could also be piscivorous have not yet made their habitat selection decision and moved. This is presumably not a significant inaccuracy because any trout large enough to be piscivorous is very unlikely to be at risk of being eaten by another trout (see "Fish length," below). Hence, when any trout with length greater than the value of *fishPiscivoryLength* executes its habitat selection, it calculates the aquatic predation risk it would be exposed to in all cells as if the density of trout predators is zero. (As discussed above, this is not necessarily equivalent to assuming survival of aquatic predation is close to 1.0.) If two or more species that have different values of *fishPiscivoryLength* are simulated, any trout with length greater than the lowest value of *fishPiscivoryLength*, no matter its species, makes habitat selection decisions as if piscivorous fish density is zero everywhere. (This assumption may deserve reconsideration when InSTREAM is applied to systems where a very broad range of trout sizes occurs and where piscivory by trout appears important.)

Depth—Aquatic predation survival is assumed to be high in water shallow enough to physically exclude large fish, or shallow enough to place large fish at high risk of terrestrial predation. The depth survival increase function is therefore a decreasing logistic function, with high survival at depths less than 5 cm (fig. 29).

Fish length—As fish grow, they become better able to out-swim piscivorous fish and fewer piscivorous fish are big enough to swallow them. The length survival increase function is therefore an increasing logistic function, the parameters for which depend on the size of the piscivorous fish. Keeley and Grant's (2001) empirical relation between the size of piscivorous stream trout and the size of their fish prey can guide the parameterization of this process for sites where trout are the only predators. Hyvarinen and Huusko (2006) found even very large piscivorous brown trout avoided prey less than 10 cm in length and were limited to prey less than 40 percent of their own length. Figure 30 illustrates a reasonable function for sites where 25- to 30-cm trout are the only piscivorous fish.

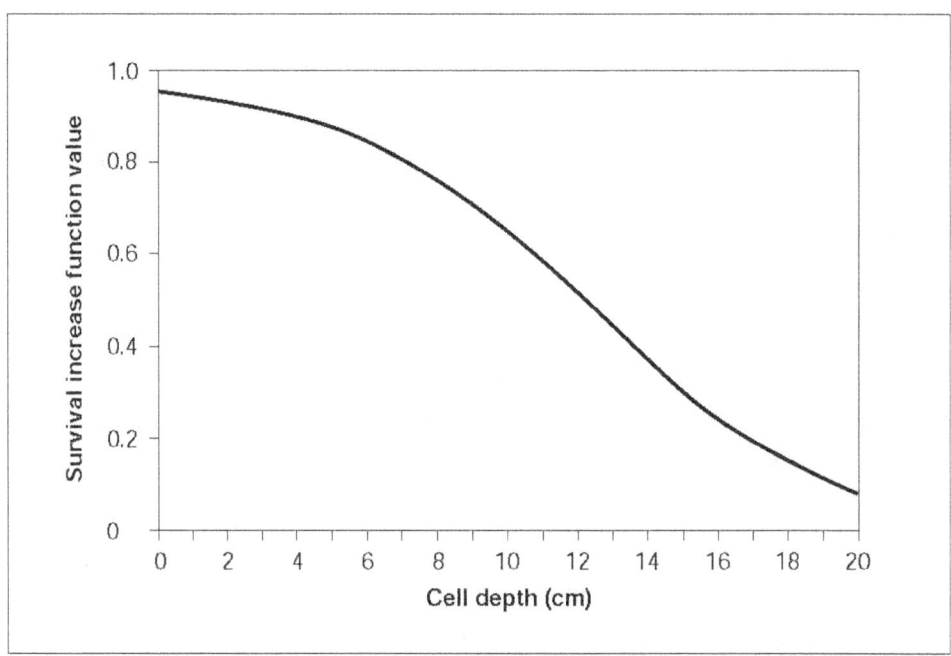

Figure 29—Depth survival increase function for aquatic predation survival.

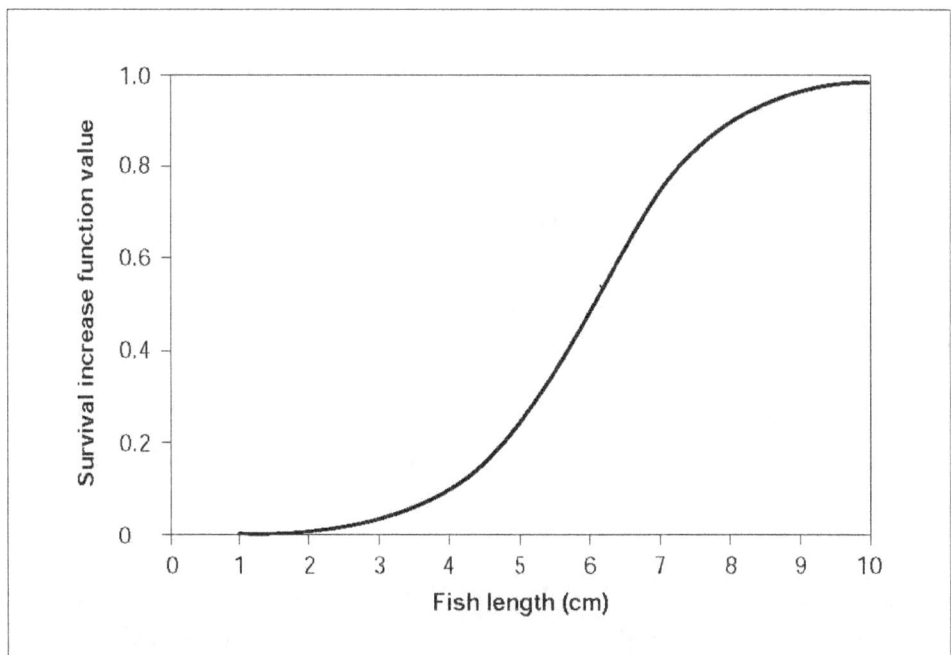

Figure 30—Fish length survival increase function for aquatic predation survival.

Feeding time—This survival increase function is the same for aquatic predation as it is for terrestrial predation. The survival increase is a decreasing logistic function of *feedTime*, the number of hours per day spent foraging. Separate parameters control the feeding time function for aquatic versus terrestrial predation, but the values recommended above for terrestrial predation are also recommended for aquatic predation.

Low temperature—This survival increase function reflects how low temperatures reduce the metabolic demands and, therefore, feeding activity of piscivorous fish. The function is based on the bioenergetics of the trout predators, using a decreasing logistic function (fig. 31) that approximates the decline in maximum food consumption (*cMax*) with declining temperature (section 6.3.5).

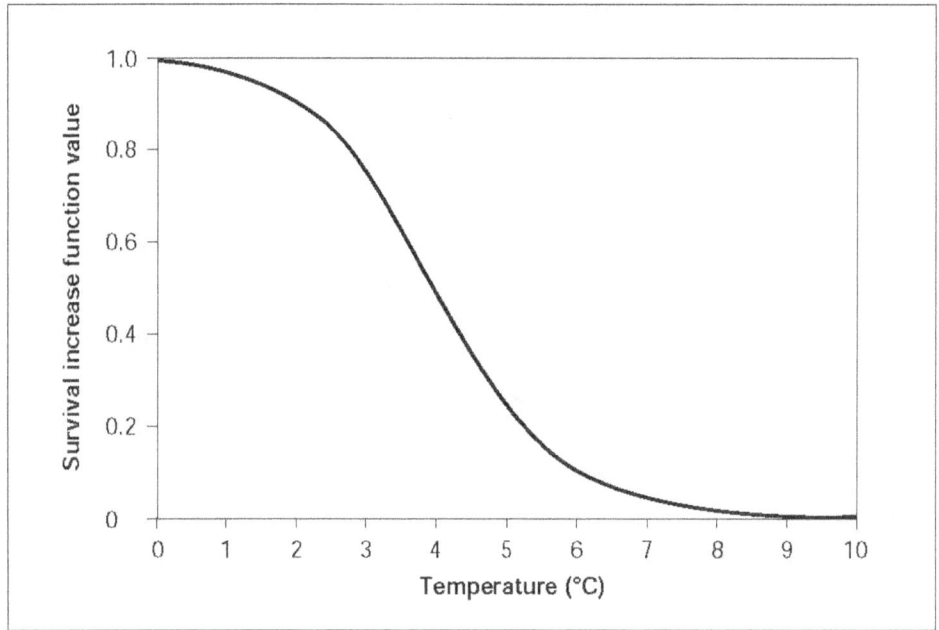

Figure 31—Temperature survival increase function for aquatic predation survival.

The parameters for the low-temperature function could be revised if aquatic predation is dominated by nontrout piscivores that do not function as well as trout at low temperatures. Parameter values could be chosen to reflect how metabolic rates and swimming performance of a less cold-adapted predator drops at temperatures below 10 °C.

Turbidity—The survival increase function for turbidity represents how encounter rates between predator and prey fish decline as turbidity increases. The turbidity function is based on experimental observations and citations provided by Gregory and Levings (1999), who found much lower rates of piscivory in a turbid river versus an adjacent clear river. Turbidity appears to reduce the ability of piscivorous fish to detect prey fish and thus the encounter rate between predator and prey (DeRobertis et al. 2003, Gregory and Levings 1999, Vogel and Beauchamp 1999). One mechanism that can offset this reduced encounter rate is that turbidity also reduces the vulnerability of piscivorous fish to terrestrial predation, making them more likely to forage in shallow habitat where small fish are likely to be found (Vogel and Beauchamp 1999). The parameters for this function provide no protection from aquatic predation at low turbidities and a 50-percent reduction in risk at 40 NTU (fig. 32). As turbidity continues to increase toward extreme values, aquatic predation risk continues to decrease but is not eliminated.

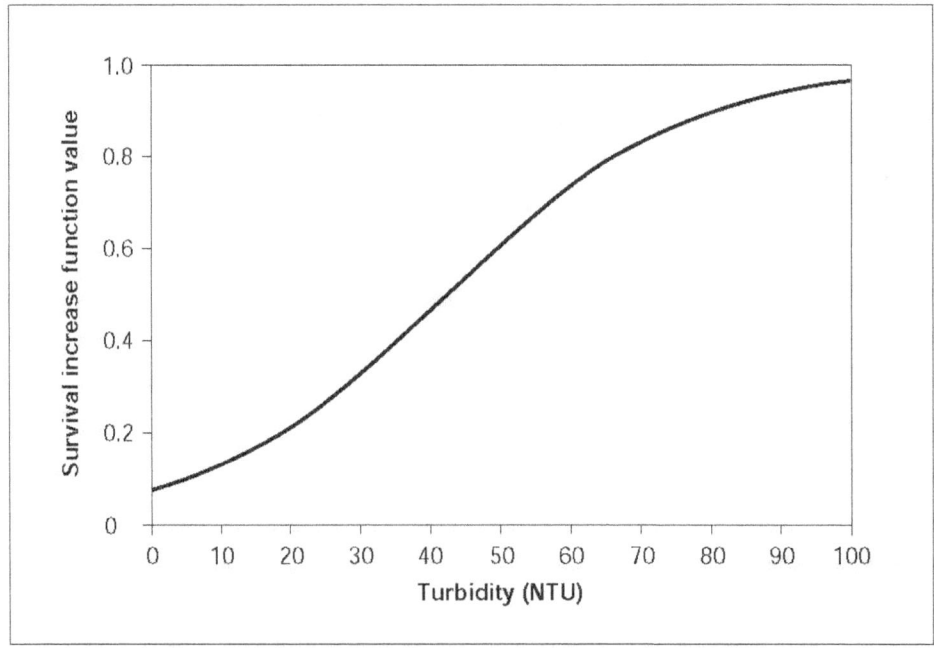

Figure 32—Turbidity survival increase function for aquatic predation survival.

6.4.7. Total survival: parameter estimation and effects of fish size, depth, and velocity—

The total survival probability for a fish in a particular cell is calculated by multiplying together the probabilities of surviving separate mortality risks. Figures 33 through 36 illustrate the variation in total survival with fish size, depth, and velocity under a certain set of conditions. They were created by plotting the total daily survival probability for four sizes of trout that all have a condition factor of 1.0, are at a temperature of 15 °C, feed for 16 h/d, have a minimum survival probability for both terrestrial and aquatic predation (*fishTerrPredMin*, *fishAqPredMin*) of 0.99, and have the values listed above for other parameters. Turbidity and distance to hiding cover were assumed to have no effect on survival, and the density of piscivorous fish was relatively high: 1 piscivore per 5 m^2 (density = 2.0×10^{-5}). Fish were assumed not to be using velocity shelters.

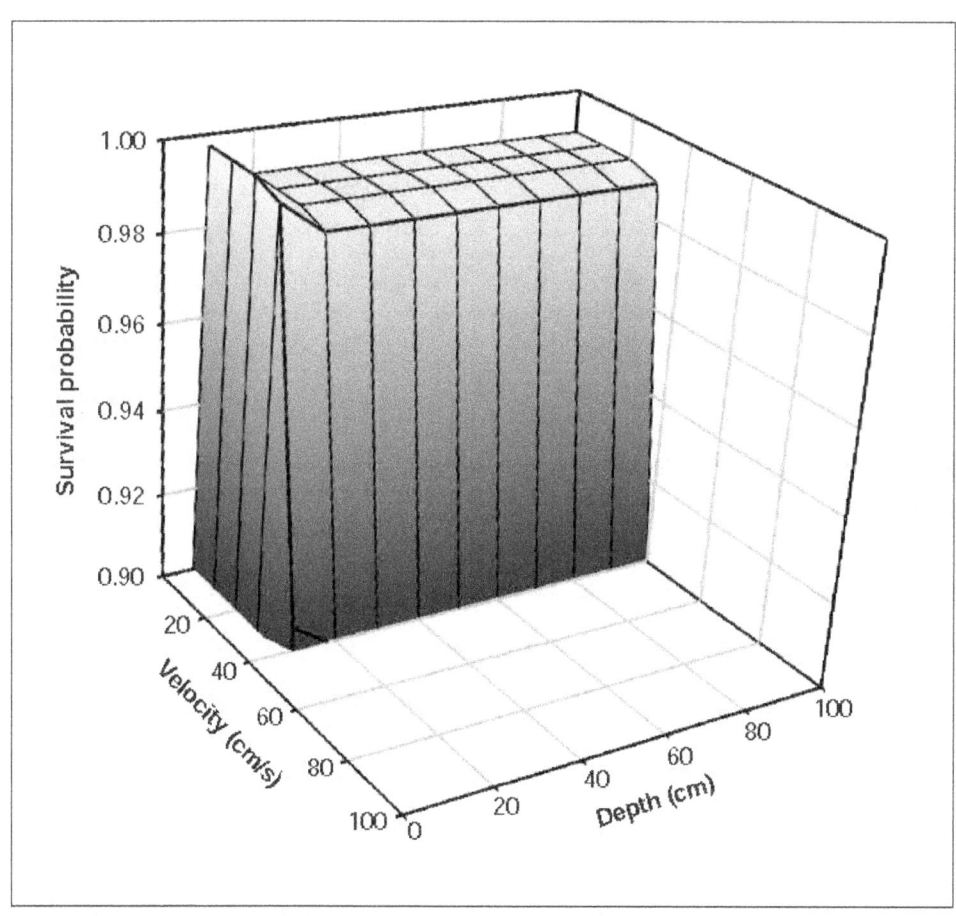

Figure 33—Total survival probability as a function of depth and velocity, 3-cm trout.

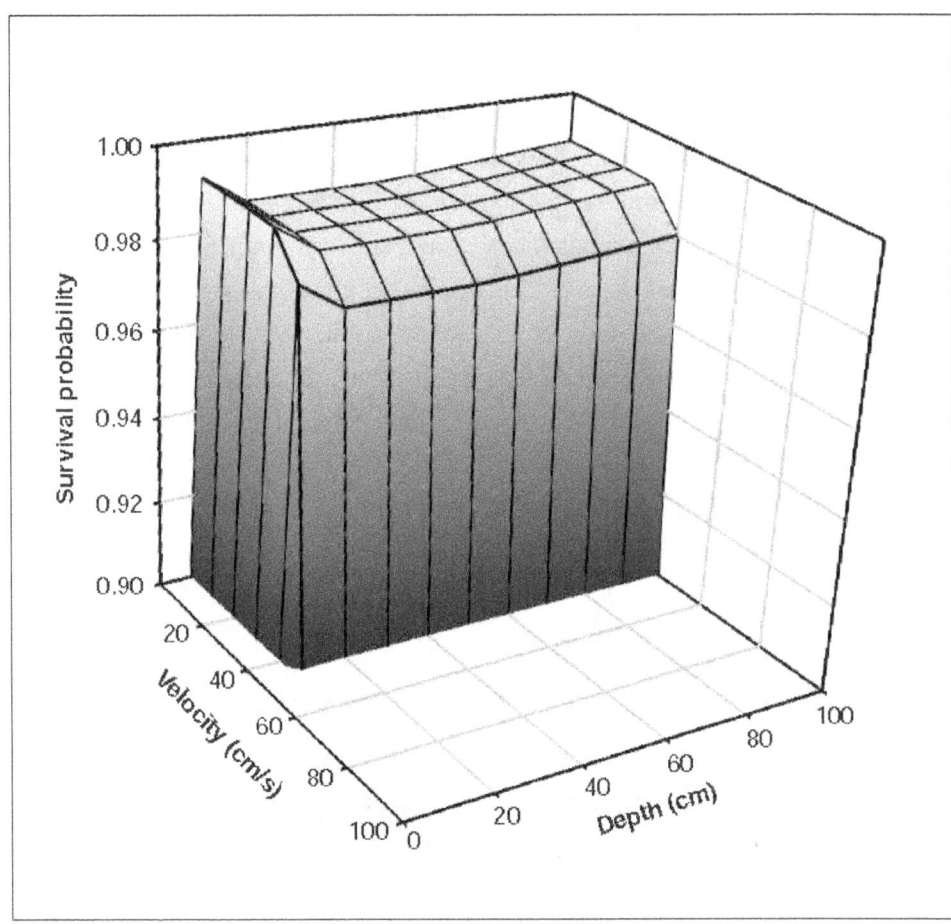

Figure 34—Total survival probability as a function of depth and velocity, 5-cm trout.

Predation survival parameters for small streams were used. The daily survival is shown on a scale of 0.95 to 1.0 because survival probabilities below 0.95 result in high mortality over several days.

The 3-cm trout (fig. 33) are vulnerable mainly to aquatic predators, as evidenced by the peak in their survival probability at depths of around 10 cm. The 5-cm trout (fig. 34) are vulnerable to both aquatic and terrestrial predators, which results in relatively low survival probabilities compared to both smaller and larger fish, while 10-cm (fig. 35) and 20-cm (fig. 36) trout are vulnerable mainly to terrestrial predators. Above 5 cm, the range of habitat conditions providing high survival increases with fish size.

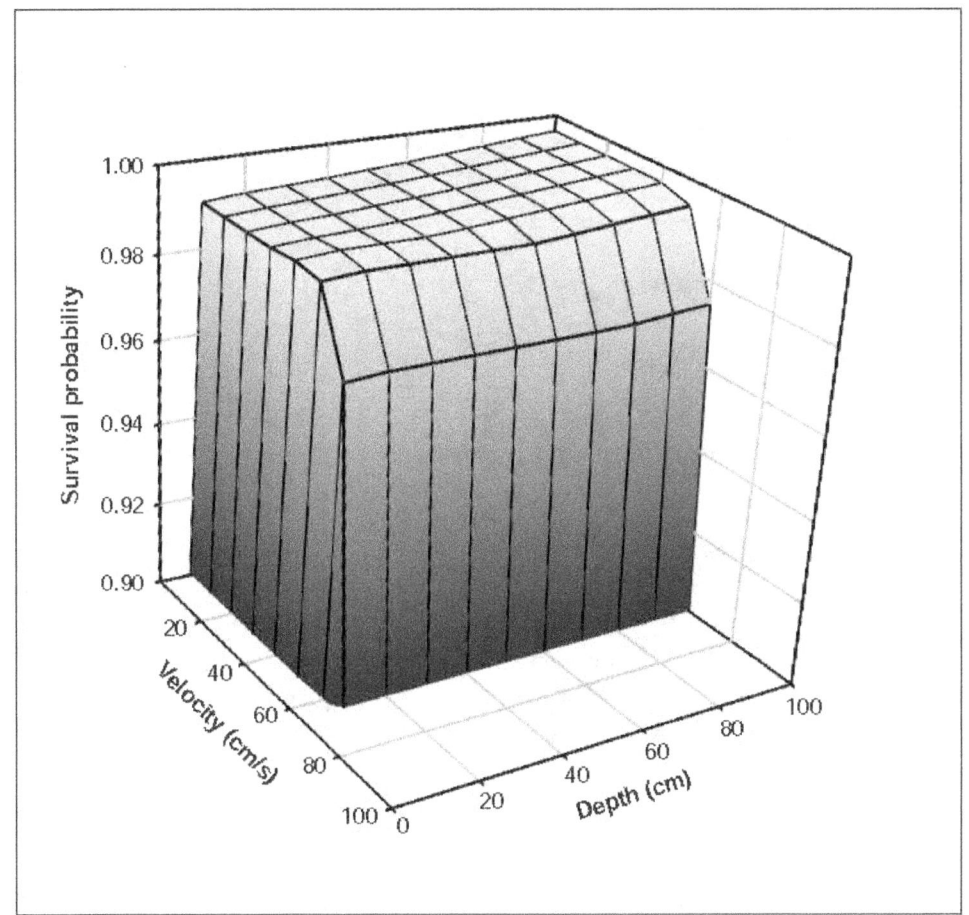

Figure 35—Total survival probability as a function of depth and velocity, 10-cm trout.

These survival probabilities are not the only processes affecting mortality rates in the modeled trout populations. The number of fish that die is also a function of the feeding and growth formulation and food availability, and of fish density. Food intake affects poor-condition mortality and habitat selection, and because survival probabilities vary with habitat, habitat selection has a major effect on survival. For example, if food is scarce (perhaps because trout abundance is high) model trout will use habitat where more food is available even if predation risk is high and pre-dation mortality will therefore increase. As a consequence of these complex interactions, mortality parameter values cannot be estimated well except by calibrating the full model as discussed in chapter 3.

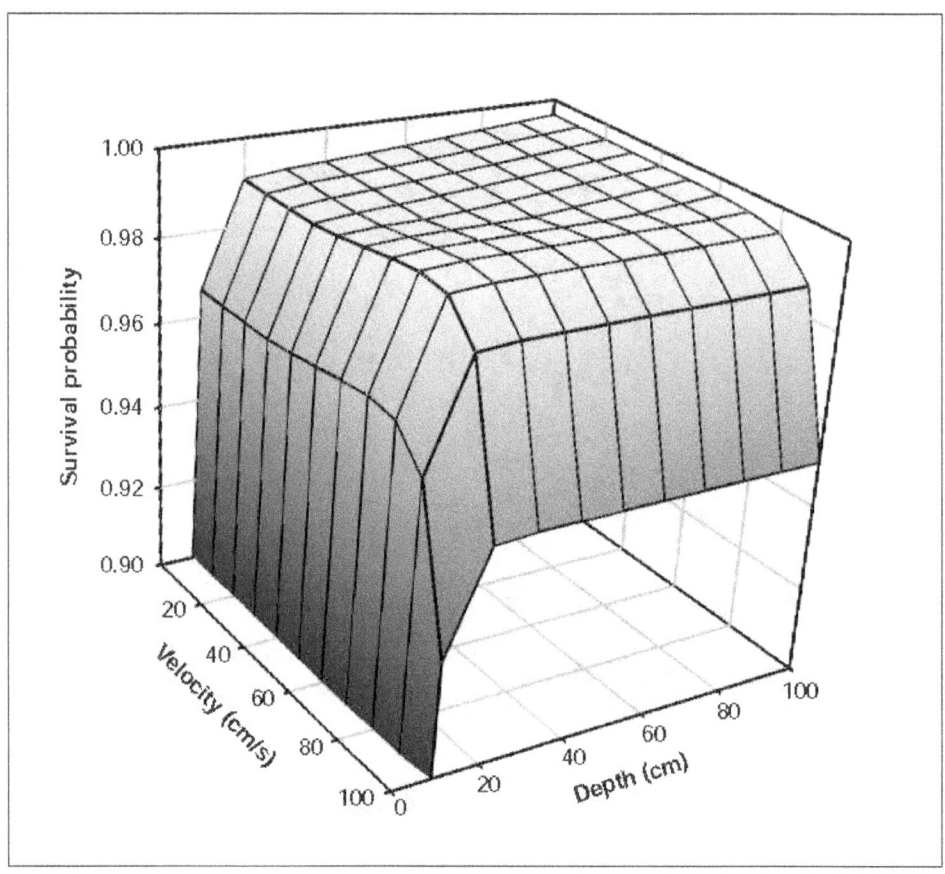

Figure 36—Total survival probability as a function of depth and velocity, 20-cm trout.

6.4.8. Demonic intrusion: experimenter-induced mortality—

The graphical interface of InSTREAM's software allows the user to select and remove individual trout from the simulation. This capability can be useful for conducting controlled simulation experiments. For example, Railsback and Harvey 2002 used it to look at how habitat use changed in a hierarchy of adult trout as the largest individuals were removed. Fish killed by the experimenter in this way are labeled as having died of "demonic intrusion," a term loosely borrowed from Hurlbert (1984).

7. Redds

Redds are the nests laid by spawning trout. When they spawn, females typically dig one or several holes in streambed gravel, deposit their eggs in the holes, and cover them. The eggs incubate in the redd until they hatch into new trout, which "emerge" by working their way up through the gravel. In InSTREAM, a redd and the eggs it contains are modeled as one object; individual fish are not tracked until they emerge. The model keeps track of the number of eggs remaining alive in each redd and determines when individuals emerge. The species of a redd and its initial number of eggs are determined by the size of the female spawner that created the redd (section 6.1.3).

Because of its objectives as a model to test the effects of flow and temperature, InSTREAM models redds with relatively little biological detail but with substantial detail in how streamflow and temperature affect egg incubation and survival. Flow and temperature can substantially affect redd success (e.g., Underwood and Bennett 1992), but other processes that may also be important (see, e.g., Groot and Margolis 1991) are not considered explicitly in InSTREAM:

- Some eggs may be diseased, unspawned, unfertilized, or washed out of the redd during its construction.
- Eggs can be killed by a variety of predators and parasites.
- Gravel size, fine sediment, and water quality can affect egg survival and development rates. In particular, low flow of water through the redd can allow metabolic wastes to accumulate and kill eggs. Deposition of fine sediment can prevent newly hatched fish from emerging.
- Salmonids go through several life stage transformations within redds. The most important of these is the transformation from eggs into alevins, which have respiratory and movement capabilities.

Redds are modeled using the following four daily actions. Scheduling of these actions is discussed in section 12.

7.1. Survival

In InSTREAM, eggs incubating in a redd are subject to five mortality sources: low and high temperatures, scouring by high flows, dewatering, and superimposition (having another redd laid on top of an existing one). Redd survival is modeled using redd "survival functions," which determine, for each redd on each day, the probability of each egg surviving one particular kind of mortality. Then, a random draw is made on a binomial distribution to determine how many eggs survive each redd mortality source. A binomial distribution is a statistical model of the (integer) number of occurrences of some event within a specified number of trials, when the probability of occurrence per trial is known. In this case, the event is death of one egg, the number of trials is the number of eggs in the redd, and the probability of occurrence is 1 minus the survival function value. Hence, the binomial distribution returns a randomly drawn number of eggs that die, given the number of live eggs and the per-egg mortality probability. (The alternative approach of multiplying the mortality probability by the number of live eggs may appear simpler, but introduces a number of numerical difficulties when the number of live eggs is small.)

The separate redd mortality sources are executed sequentially: the eggs killed by one source are subtracted from the number alive before the next source is processed. The order in which redd survival functions are evaluated is defined in section 12.3.

The kinds of mortality represented, and the survival function methods, were selected considering that the objectives of InSTREAM focus on flow and temperature effects on trout populations. For example, there is no redd survival function related to spawning gravel quality. Spawning gravel quality has several effects on redd success (Kondolf 2000), but InSTREAM is not designed to address these. (The spawning site selection criteria [section 6.1.2] allow a fish to spawn in a cell that has little or no gravel; there is no redd mortality penalty for doing so. The exception is that if superimposition occurs in a cell with little spawning gravel [unlikely unless gravel is rare] then superimposition mortality is likely to be high.) For several of the redd mortality sources (especially dewatering and superimposition), more detailed and mechanistic approaches are available in the literature and could be added to InSTREAM in situations where these mortality sources are believed to be important.

7.1.1. Dewatering—

Dewatering mortality occurs when decreases in flow expose redds: eggs can be killed by dessication or the buildup of waste products that are no longer flushed away. Reiser and White (1983) did not observe significant mortality of eggs when water levels were reduced to 10 cm below the egg pocket for several weeks. However, they also cited literature indicating high mortality when eggs and alevins are only slightly submerged (which may yield poorer chemical conditions than being dewatered), and high mortality for dewatered alevins. Because InSTREAM does not distinguish between eggs and alevins, these processes are not modeled mechanistically or in detail. The dewatering survival function is simply that if depth is zero then the daily fraction of eggs surviving is equal to the fish parameter *mortReddDewaterSurv*. This parameter has a suggested value of 0.9, which reflects the variability in dewatering effects. Egg survival may be high when a redd is first dewatered, which suggests caution using lower values of *mortReddDewaterSurv*.

7.1.2. Scouring and deposition—

Scouring and deposition mortality results from high flows disturbing the gravel containing a redd. Low survival can be expected for eggs scoured out of a redd. Deposition of new gravel on top of a redd may make water flowing through the redd inadequate to transport oxygen and waste materials, or may prevent emergence of fish. Deposition is especially likely to reduce survival if it includes fine sediment. This redd mortality source can be very important to trout populations and communities.

Although empirical methods allow prediction of the potential for scouring as a function of shear stress and substrate particle size at the reach scale, scour and deposition at the scale of individual redds is difficult to predict (Haschenburger 1999, Wilcock et al. 1996) and probably best represented as stochastic. Consequently, InSTREAM adopts an approach for predicting the probability of redd scouring or deposition from the empirical, reach-scale work of Haschenburger (1999). This approach was developed for bar-pool, gravel-bed channels and may not be appropriate for sites where spawning gravels occur mainly in pockets behind obstructions (where scouring may be even less predictable). InSTREAM should be considered substantially more uncertain for sites where populations are strongly limited by redd scour, especially if spawning is limited to pocket gravels. However, all models of trout populations or habitat are likely less useful at such sites.

Haschenburger (1999) observed the spatial distribution and depth of scour and deposition at a number of flow peaks in several study sites in gravel-bed rivers. The proportion of a stream reach that scoured or filled to a specified depth during a high-flow event was found to follow an exponential distribution, the parameter for which (*scourParam*) varies with site-average dimensionless (Shields) shear stress. Therefore, InSTREAM assumes the probability of a redd being destroyed is equal to the proportion of the stream reach scouring or filling to depths greater than the value of the fish parameter *mortReddScourDepth* (cm). Consequently, the probability of a redd **not** being destroyed (*scourSurvival*) is equal to the proportion of the stream scouring or filling to a depth **less than** the value of *mortReddScourDepth*. This scour survival probability is estimated from the exponential distribution model of Haschenburger (1999); the proportion of the stream scouring to less than a given depth is the integral of the exponential distribution between zero and the depth:

$$scourSurvival = 1 - e^{\,scourParam \times mortReddScourDepth}$$

(The value of *scourSurvival* is set to 1.0 if *scourParam* × *mortReddScourDepth* is greater than 100. This allows users to effectively turn scouring and deposition mortality off by using a very large value of *mortReddScourDepth*, e.g., 10,000 cm.

The value of *scourParam* was modeled by Haschenburger empirically:

$$scourParam = 3.33 \times e^{-1.52 \times (shearStress/0.045)}$$

where *shearStress* is the peak Shields stress (measured at a reach scale) occurring during the high-flow event. Shields stress is a dimensionless indicator of scour potential often used in modeling sediment transport (Yalin, 1977). Shields stress increases with flow, a relationship represented in InSTREAM by the equation:

$$shearStress = habShearParamA \times flow^{habShearParamB}$$

where *habShearParamA* (s/m^3) and *habShearParamB* (unitless) are habitat reach parameters. These are habitat parameters because they are highly specific to each reach. Methods for estimating *habShearParamA* and *habShearParamB* are discussed in chapter 3.

The fish parameter *mortReddScourDepth* can be evaluated as the egg burial depth, the distance down from the gravel surface to the top of a redd's egg pocket. Scour to this depth is almost certain to flush eggs out of the redd. Deposition of new material to this depth over the redd would double the depth to the egg pocket and severely reduce the survival and emergence of its embryos. Egg-pocket depths of 5 to 10 cm are reasonable for small trout using relatively small gravel (DeVries 1997).

As an example of the influence of scour in the model, scour survival parameters for the Little Jones Creek study site (*habShearParamA* = 0.019, *habShearParamB* = 0.383, *mortReddScourDepth* = 5, 10 cm) produce the decreasing relation between peak flow and survival of redd scouring in figure 37. As this figure illustrates, redd survival of scour can be quite sensitive to the value of *mortReddScourDepth*.

This model of scour estimates the probability of a redd surviving scour in each high-flow event, rather than in each daily time step. The single survival probability is applied to all redds, assuming that if scouring occurs no eggs survive. (It is important to note that InSTREAM calculates scouring survival from mean daily flows, whereas Haschenburger [1999] based her model on instantaneous peak flows. This approximation is made to avoid needing to input daily peak flows, but will cause scouring mortality to be underestimated when runoff is rapid.) The following steps are used for each redd, on each day:

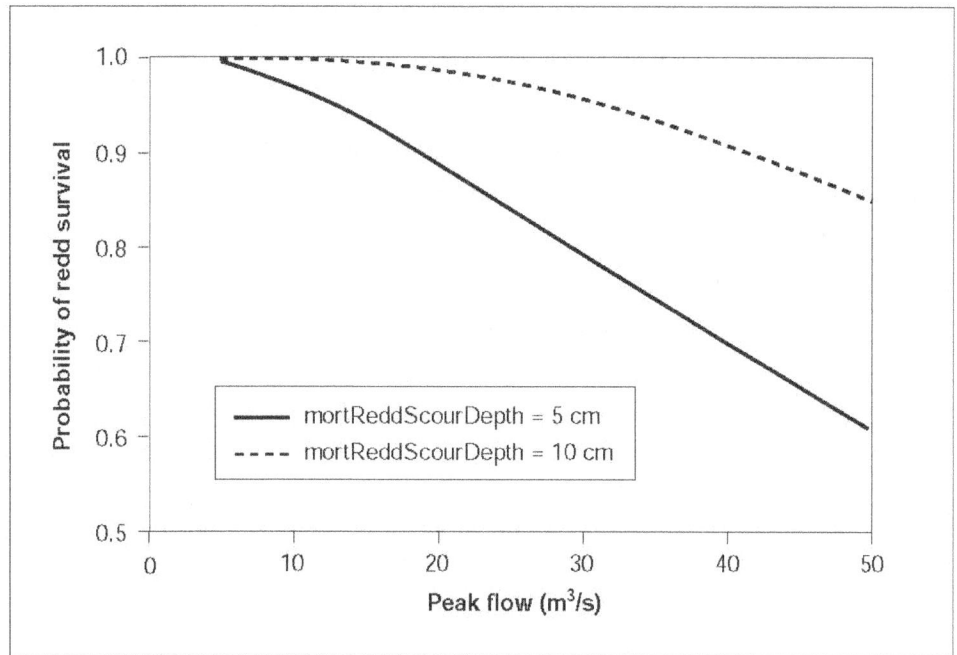

Figure 37—Example redd scour and fill survival function. The Y axis is the probability of the redd not being destroyed during a peak flow event.

- Determine whether the current day's flow in the redd's reach is greater than both the previous day's and the following day's flow. If so, then the following steps are conducted. If not, then all eggs survive.

- Calculate the value of *scourSurvival*, using the above equations and the current day's flow.

- Draw a uniform random number between 0 and 1. If its value exceeds the value of *scourSurvival*, no eggs survive. Otherwise, the fraction of eggs surviving is 1.0.

To avoid the need for flow data for the date preceding the start of a model run, redd scour is not executed on the first day of a run. However, redd scour can be executed on the last day, so flow input must extend at least 1 day past the last simulation date.

7.1.3. Low temperature—

Both low and high temperatures cause mortality in eggs, at temperatures much different than those causing mortality in fish. Mortality from high and low temperatures is modeled separately. Logistic functions represent the available data well.

The daily fraction of eggs surviving low temperatures is modeled as an increasing logistic function of temperature. Parameter values appear to differ among salmonids, especially between fall versus spring spawners. In developing parameter values from published data on egg survival, it is important to remember that eggs incubate slowly at low temperatures, so even apparently high daily survival rates can result in low egg survival over the entire incubation period.

Parameter values for spring-spawning rainbow trout and fall-spawning brown trout (table 19, fig. 38) have been determined from data compiled by Brown (1974). These data indicate that rainbow trout spawn at temperatures as low as 3 to 5 °C, and eggs have a 90-percent survival rate over a 100-d incubation period at 3 °C (daily egg survival = 0.999). We assume a daily survival rate of 0.9 (very low long-term survival) for 0 °C. Brown trout egg incubation can take over 150 days at very low temperatures (Brown 1974). Parameter values for brown trout were estimated by assuming 90-percent egg survival over 150 days at 1 °C (daily survival of 0.9993) and daily survival of 0.9 at 0 °C.

Table 19—Parameter values for low-temperature redd mortality

Parameter	Definition	Species	Value
mortReddLoTT1	Temperature at which low-temperature survival is 10 percent (°C)	Rainbow Brown	-3 -0.8
mortReddLoTT9	Temperature at which low-temperature survival is 90 percent (°C)	Rainbow Brown	0 0

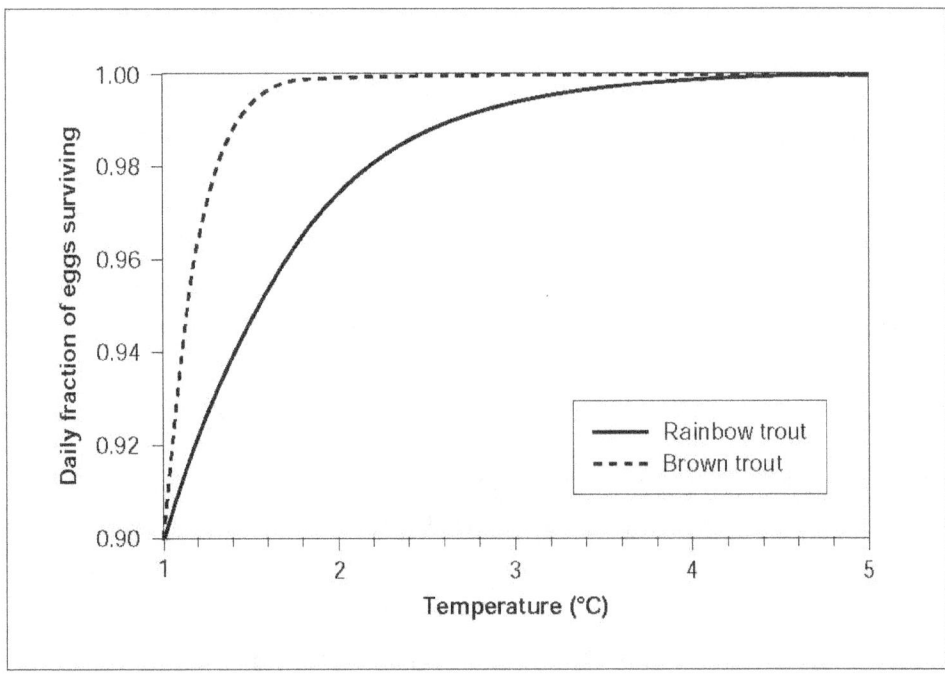

Figure 38—Low-temperature redd survival function, for rainbow and brown trout parameter values.

7.1.4. High temperature—

High temperatures can induce direct mortality in trout eggs, and also promote fungus and disease. The fraction of eggs surviving high temperatures is modeled as a decreasing logistic function of temperature (fig. 39). Parameter values for rainbow trout (also used for cutthroat trout by Railsback and Harvey 2002) are based on interim results of lab studies conducted by the University of California at Davis (Myrick 1998). These data showed daily survival rates declining from about 0.9998 at 11 °C to about 0.985 at 19 °C. The resulting parameter values (table 20) appear to indicate high survival at high temperatures, but low survival if temperatures are elevated for long periods. Fall spawning trout are likely to be less adapted to high incubation temperatures. Parameter values for brown trout in table 20 were arbitrarily set to 5 °C less than the rainbow trout values and should not be considered reliable.

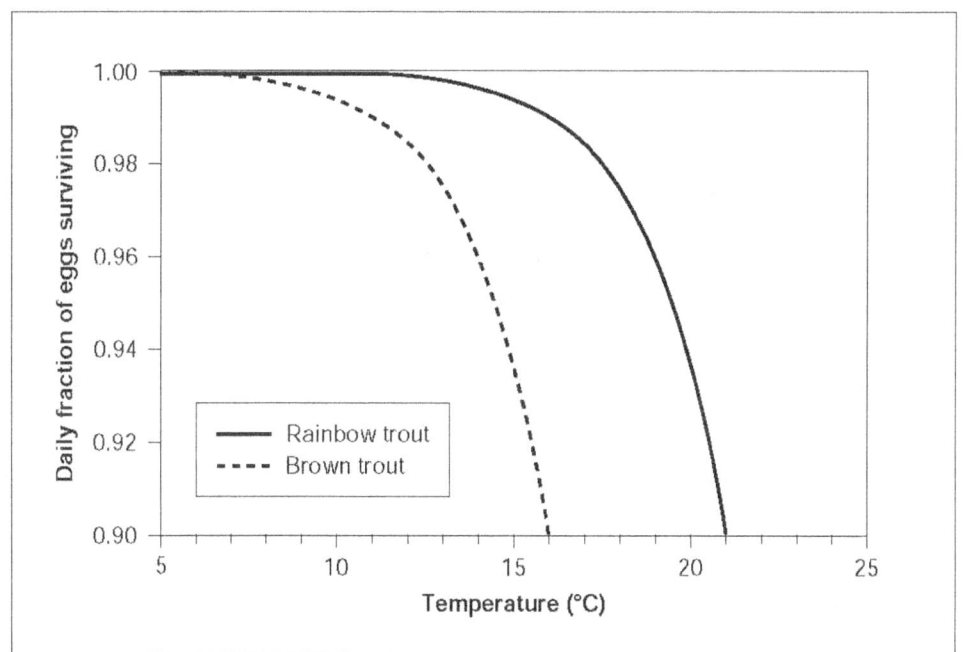

Figure 39—High-temperature redd survival function, for rainbow and brown trout parameter values.

Table 20—Parameter values for high-temperature redd mortality

Parameter	Definition	Species	Value
mortReddHiTT1	Temperature at which high-temperature survival is 10 percent (°C)	Rainbow Brown	30 25
mortReddHiTT9	Temperature at which high-temperature survival is 90 percent (°C)	Rainbow Brown	21 16

7.1.5. Superimposition—

Superimposition mortality can occur when a new redd is laid over an existing one. Females digging new redds can disturb existing redds and cause egg mortality through mechanical damage or by displacing eggs from the redd environment. Superimposition typically kills many but not all eggs in a redd (Essington et al. 2000, Hendry et al. 2003). For simplicity, InSTREAM currently assumes that superimposition is accidental with no bias for or against spawning over existing redds. Stream trout may indeed intentionally superimpose their redds over existing ones (Essington et al. 1998), a practice that has the advantages of reducing (a) the work necessary to clean redd gravels and (b) the competition that the spawner's offspring will face (Morbey and Ydenberg 2003). The formulation could be modified to represent intentional superimposition and the complex effects that it might have, but there is currently little known about factors influencing intentional super-imposition.

Superimposition mortality is modeled as a function of the area disturbed in creating the new redd and the area of spawning gravel available. The model subjects each redd to these steps at daily intervals:

1. Determine if one or more new redds were created in the same cell on the current day. If not, then superimposition survival is 1.0.

2. If one or more redds (of any species) were created in the same cell, the probability of each new redd causing superimposition (*reddSuperImpRisk*, unitless) is equal to the area of a redd (*reddSize*, cm^2, a fish parameter that can be species-specific) divided by the area of spawning gravel in the redd.

$$reddSuperImpRisk = \frac{reddSize}{(cellArea \times cellFracSpawn)}$$

3. A random number is drawn from a uniform distribution between 0 and 1; if it is less than *reddSuperImpRisk*, then superimposition mortality occurs.

4. If superimposition mortality occurs, then the fraction of eggs surviving is the value of another random number drawn from a uniform distribution between 0 and 1.

5. Steps 2 through 4 are executed once for each new redd placed in the cell on the current day.

Note that the value of *reddSuperImpRisk* can be greater than 1.0 if *cellFrac-Spawn* is very small; in that case, superimposition egg mortality always occurs. In the event that *cellFracSpawn* is zero, there is no risk of superimposition. This assumption is made because there is no gravel to be disturbed by another spawner.

Because of how the parameter *reddSize* is used in this formulation, it is defined as the area a spawner disturbs in creating a new redd. Field observations suggest a *reddSize* value of 1200 cm^2 (the area of a circle with a diameter of 35 cm) for relatively small trout.

7.2. Development

To predict the timing of emergence, the developmental status of a redd's eggs is updated daily using the fractional development approach of Van Winkle et al. (1996). This approach is based on accumulated degree-days, a simple and reasonably accurate technique for modeling incubation. (Beer (1999) reviewed alternative models of salmonid egg development.)

Model redds accumulate the fractional development that occurs each day (*reddDailyDevel*), a function of temperature. This means the redd has a variable *fracDeveloped* that starts at zero when the redd is created and is increased each day by the value of *reddDailyDevel*. The daily value of *reddDailyDevel* is determined using the equation:

$$reddDailyDevel = reddDevelParamA + (reddDevelParamB \times temperature) + (reddDevelParamC \times temperature^2)$$

The parameters for this equation probably differ among species and among populations that spawn at different times of year. Parameter values for spring-spawning rainbow trout and fall-spawning brown trout were developed by Van Winkle et al. (1996) (table 21). Railsback and Harvey (2001) found the rainbow trout parameters in table 21 yielded reasonable results for a cutthroat trout population in coastal California.

Table 21—Parameter values for egg development rates

Parameter	Definition	Rainbow, cutthroat trout value (spring spawning)	Brown trout value (fall spawning)
reddDevelParamA	Constant in daily redd development equation (unitless)	-0.000253	0.00313
reddDevelParamB	Temperature coefficient in daily redd development equation (°C^{-1})	0.00134	0.0000307
reddDevelParamC	Temperature squared coefficient in daily redd development equation (°C^{-2})	0.0000321	0.0000934

7.3. Emergence

"Emergence" is the conversion of each surviving egg into a new trout object. When a redd's value of *fracDeveloped* equals or exceeds 1.0, fish can begin to emerge. New fish emerge over several days.

7.3.1. Emergence timing—

New fish emerge over several days beginning on the day when *fracDeveloped* reaches 1.0. Modeling emergence to occur over several days reproduces observed natural variation in emergence timing and can potentially have strong effects on survival of newly emerged trout. These fish compete with each other for food as soon as they emerge. If all emerged on the same day, without time for some to move, competition would probably be overestimated. As a simple way to spread emergence over several days, InSTREAM assumes that 10 percent of the redd's fish emerge on the first day of emergence; 20 percent of the remaining fish emerge on the next day; 30 percent of the remaining fish emerge on the third day; etc, until 100 percent of remaining fish emerge on the 10th day. For example, if a redd contains 100 fish on the day that development is complete, 10 new free-swimming trout will be created on that day and 90 fish will remain. On the next day (assuming no mortality occurs), 18 new free-swimming individuals will be created (20 percent of 90) and 72 fish (= 90 - 18) remain in the redd. On the third day of emergence, 21 fish (30 percent of 72, truncated to an integer) emerge. As emergence proceeds, the fish remaining in a redd remain susceptible to egg mortality.

7.3.2. New fish attributes—

For each new fish that emerges, the model assigns these attributes:

- New fish are assigned the species of the spawner.
- The fish's location is the same habitat cell as its redd.
- Gender is assigned randomly, with equal probability of being male or female.
- The length of each individual fish (*fishLength*, cm) is assigned from a random normal distribution with mean equal to the fish parameter *reddNewLengthMean* (cm) and standard deviation equal to the parameter *reddNewLengthStdDev* (cm). However, no fish are given lengths less than half the mean: if the randomly drawn length is less than half the value of *reddNewLengthMean*, a new length is drawn.
- Weight (*fishWeight*, g) is calculated from length, using the length-weight relationship and parameters used in modeling growth (section 6.3.1) and to create initial fish (section 8.2). Fish are assumed to have a normal condition factor (*fishCondition* = 1.0) when they emerge:

$$fishWeight = fishWeightParamA \times fishLength^{fishWeightParamB}$$

Variation among individuals in length at emergence is represented because habitat selection (and, consequently, growth and survival) is modeled using a length-based hierarchy (section 6.2.1). Elliott (1994) found that fish emerging from a redd only slightly differ in size, but the variation gives larger fish an advantage in dominance likely to persist and grow over time because competition among newly emerged trout is often intense (a process captured by InSTREAM) (Railsback et al. 2002).

Table 22 provides example length parameters for newly emerged fish from a study of coastal cutthroat trout in Washington (June 1981). This study measured lengths of newly emerged fry found in a downstream trap. A few of these fry had lengths between 2.4 and 2.7 cm, but most were between 2.7 and 3.0 cm. Elliott (1994) observed a coefficient of variation of 0.07 in length at emergence for brown trout at several sites. This value is converted to the standard deviation in length (with a coefficient of variation of 0.07 and a mean length of 2.8 cm, *reddNew-LengthStd* is 0.2 cm). Parameter values for other species are likely to be available from the literature or from hatchery data.

Table 22—Parameter values for size of newly emerged fish

Parameter	Definition	Cutthroat trout value
reddNewLengthMean	Constant for new fish length equation (cm)	2.8
reddNewLengthStdDev	Standard deviation in length of newly emerged fish (cm)	0.2

The previous model of Van Winkle et al. (1996) assumed that bigger spawning females produce bigger eggs, and that bigger eggs produce bigger fish at emergence. This effect of spawner size on offspring size may be important for salmon and large trout where variation in spawner size is large. It is also a mechanism making the offspring of larger fish more likely to be successful. However, relationships among sizes of spawners, eggs, and emergent fish are inconsistent and not well known for most populations. This mechanism does not appear important for the objectives of InSTREAM.

7.4. Empty Redds

As described in the previous sections, the number of eggs remaining in redds is reduced when eggs die or fish emerge. When the number of remaining eggs in a redd reaches zero, the redd is dropped from the model.

8. Initialization

This section describes the methods used to initialize the habitat and fish populations for each new model run. Although this section mentions some of the input types and files, chapter 4 provides complete documentation of file and input types.

8.1. Habitat Initialization

A model run starts by reading in the habitat characteristics that do not change during the simulation. These characteristics are the number of reaches and arrangement of reaches, the location and dimensions of cells in each reach, the values of cell variables that do not change with time, and the lookup tables used to calculate daily depth and velocity in each cell (section 5.2). Finally, variables that depend on time-series input (reach temperature, flow, turbidity; cell depth and velocity) are initialized with the input data for the first simulation date.

8.2. Fish Initialization

The initial fish population is built from input data giving the initial abundance, mean length, and standard deviation in length for each age class of each species. (Age classes are defined in chapter 1.) Separate fish initialization data are provided for each habitat reach.

The methods used to initialize fish are the same as those used to create new fish from redds (section 7.3.2). The length of each fish is drawn randomly from a normal distribution with age-specific means and standard deviations from the initial population data file. (Ideally, these data are based on empirical observations.) The lengths of initial fish are restricted to being greater than half the mean length for their age class. Weights are calculated from length using parameters *fishWeightParamA* and *fishWeightParamB*.

Each fish's location is assigned stochastically while avoiding extremely risky habitat. Initial fish are distributed randomly, after which the first day's habitat selection action lets the fish move to more suitable habitat. This approach is designed to be simple and avoid bias in initial locations. However, the method also limits the random distribution of fish to cells where the fish are not immediately at high risk of mortality from high velocity or stranding. Small fish especially may have a maximum movement distance (section 6.2.2) too small to allow them to find reasonably safe habitat during habitat selection on the first simulation day. The following steps are used to assign a fish to its initial cell after the habitat reach has been initialized with the flow for the first simulation day.

1. A point in the habitat reach is randomly selected. This point's X coordinate is drawn randomly between zero and the reach's length, and its Y coordinate is drawn between zero and the width of the reach's widest transect.

2. The cell containing the random point is identified. If the point is not within a cell, then Step 1 is repeated to select a new random point.

3. If the cell has a depth of zero, then steps 1 and 2 are repeated to identify a new cell.

4. If the cell's velocity puts the fish at extreme risk of high-velocity mortality, then steps 1 through 3 are repeated to identify a new cell. The degree of risk is determined by (a) calculating the fish's maximum swimming speed in the cell, (b) calculating the ratio of maximum swim speed to cell velocity, and (c) determining whether this ratio is greater than the parameter *mortFishVelocityV9*. If so, then the fish's daily probability of surviving high-velocity mortality is less than 90 percent, so steps 1 through 3 are repeated to select a new cell. Otherwise, the cell becomes the fish's initial location.

5. If steps 1 through 4 result in step 1 being repeated 10,000 times without locating an acceptable cell, then the high-velocity criterion (step 4) is abandoned and steps 1 through 3 are repeated.

6. Model execution terminates if step 1 is again repeated 10,000 times. If this limit is reached, it is very unlikely there are any cells with non-zero depth.

Fish have a variable *spawnedThisSeason* (section 6.1.1), which is set to NO when fish are initialized.

8.3. Redd Initialization

Redds cannot be initialized at the start of a simulation. Redds can only be created by spawning fish.

9. Random Year Shuffler

One concern in using models like InSTREAM is the sensitivity of the results to the specific years chosen for simulation. Does a simulation experiment using input from 1990–99 produce the same conclusions as an experiment using input from 1980–89? Would different conclusions be drawn if the input included more wet years and fewer dry years, or if the wet and dry years occurred in a different order? The optional year shuffler in InSTREAM can randomize the sequence of simulated years with or without replacement. The year shuffler works like a time machine, causing the model's clock to jump to a random year at the start of each simulation year. The model's clock then determines which input is used and provides the date that output is labeled with. The following steps are used.

1. The simulation period (defined by the simulation start and end dates specified by the user) is divided into simulation years. New simulation years start on the same month and day of the month as the simulation start date. A list of these simulation years is created.

2. The list of simulation years is then randomized, either with or without replacement as specified by the user.

3. Each time the model reaches the beginning of a new simulation year (the month and day of the month are equal to those of the simulate start date), the next year is taken from the randomized list.

4. When the next year is taken from the randomized list of years, the model's clock jumps to that year. The model's clock determines which input data are used, and is used to label output.

For example, a model run is set to run for exactly 5 years, from 10/1/1990 to 9/30/1995. The simulation years are then 1990–94. Using year shuffling with replacement yields: 1991, 1990, 1994, 1992, 1993. Therefore, the model starts on 10/1/1991. At the end of the first simulation year (9/30/1992), the model's clock jumps to 10/1/1990 and runs for another year until 9/30/1991. Then the clock jumps to 10/1/1994 and the model run until 9/30/1995. Finally, the clock jumps to 10/1/1992 and runs (because 1992 is followed by 1993 in the randomized order) until the simulation ends on 9/31/1994.

(The year shuffler works but is more complicated when the simulation period includes partial years. For example, a model run is set to run from 10/1/1990 to 12/31/1995. Now there are six simulation years, because the model continues to run past 10/1/1995. Year shuffling with replacement yields: 1991, 1990, 1995, 1994, 1992, 1993. Therefore, the model starts on 10/1/1991. After 9/30/1992 the model's clock jumps to 10/1/1990; after 9/30/1991 the clock jumps to 10/1/1995 and the model run until 9/30/1996. Then the clock jumps to 10/1/1994, and after 9/31/1995 it jumps to 10/1/1992 and runs until the simulation ends with the partial year 10/1/1993 to 12/31/1993. Note that using year shuffling with partial years requires the user to provide more input data than nonrandomized runs. The example model run included the period 1/1/1996 through 9/30/1996, even though the model end date was set to 12/31/1995. The year shuffler requires the user to provide input data for **all** of each simulation year.)

The methods used to determine whether a fish spawns (section 6.1.1) and whether redd scour mortality occurs (section 7.1.2) depend on the relation between streamflow on the current day and on the following day (i.e., a fish will not spawn if tomorrow's flow is much less than today's). In these methods, the value of tomorrow's flow is determined by ignoring year shuffling: tomorrow's flow is always the flow on the next calendar day even if the model's clock is about to jump to a different year.

10. Random Number Generation

InSTREAM models several processes stochastically (e.g., fish initialization, fish survival), using pseudorandom numbers to determine outcomes. The method used to generate pseudorandom numbers is an important issue for any stochastic simulation model, as poor quality or misused random number generators can bias simulation results.

All pseudorandom numbers in InSTREAM are generated by the MT19937 "Mersenne Twister" algorithm, the default generator in the Swarm software platform used to implement InSTREAM. (See SDG 2000 for additional information and references.)

The random number generator used for all stochastic processes in InSTREAM, with one exception described in the following paragraph, is initialized with a random number seed, *randGenSeed*, provided by the user as a model parameter. If two model runs use the value of *randGenSeed* **and** exactly the same input and parameters, the two runs will produce exactly the same results. However, any change to input (parameter values, input data, simulation dates, etc.) is very likely to alter the number of times the random number generator is called and, therefore, the outcome of all stochastic processes. Replicate simulations are produced by altering only the value of *randGenSeed*. (The software for InSTREAM can create replicate simulations automatically; see chapter 4.)

The optional year shuffler (section 9) uses a separate random number generator. The year randomizer uses its own generator and seed (model parameter *shuffle-YearSeed*) so that year randomization can be controlled separately. For example, multiple model runs that use the same value of *shuffleYearSeed* but different values of *randGenSeed* will produce replicate simulations that all use the same sequence of simulation years.

11. Observation and Output

Individual-based models such as InSTREAM are like real ecosystems in that our perception and understanding of them is affected by how we observe them. InSTREAM creates a complex digital world of changing habitat and variable individuals, and the conclusions drawn from simulations can depend very much on what data are collected and reported. As with real ecosystems, it is infeasible to observe everything that happens in InSTREAM, so the methods used to observe and report results must be carefully designed.

InSTREAM produces six major categories of output. The software users guide in chapter 4 documents methods for the control and interpretation of these outputs.

11.1. Graphical Displays

People are best able to absorb and interpret complex information presented visually. Therefore, InSTREAM provides a graphical display of habitat cells and the location of fish and redds as the model executes. The display also indicates the size and species of each fish. This display provides a plan view of modeled reaches.

The graphical display is most useful for understanding patterns of fish habitat use. It is the only output that provides the explicit location of individual fish. Although the graphical display produces no numerical output that can be analyzed, it is essential for developing understanding of the model, especially its habitat and habitat selection methods.

11.2. Summary Population Statistics

It would be very cumbersome and unhelpful to output the state of each individual fish over time, so instead summary statistics are generated from InSTREAM and reported via file output. These statistics include abundance, mean and maximum fish length, and mean and maximum fish weight, all broken out by species, age class, and habitat reach. The software is easily modified to obtain additional output variables or to break statistics out by additional factors.

11.3. Habitat and Habitat Use Statistics

Resource managers and researchers commonly seek to contrast habitat use with habitat availability. InSTREAM supplies this information categorically.

11.4. Fish Mortality

Understanding how many fish die of each mortality source is often important. InStream records the cause of death for each model fish and the output describes the cumulative number of fish that have died of each mortality source.

11.5. Redd Status and Mortality

Redd output reports when a redd was created, how many eggs were created, and when the redd was removed from the model because all its eggs died or emerged. Redd mortality output reports how many eggs from each redd died from each redd mortality source.

11.6. Intermediate Output

The previous five kinds of observations can be considered "final" results: they describe what happened during simulations, but not why individuals behaved as they did. Intermediate results include the state and decisions of individuals as they proceed through each day's actions. Output of intermediate results can be important for testing and understanding the model. For example, if a particular kind of fish (e.g., small juvenile trout) exhibits an unexpected behavior—using deep instead of shallow habitat—intermediate output will be needed to understand whether this unexpected behavior is due to a flaw in the habitat selection method or is simply the result of an unusual situation (e.g., a lack of hiding cover in shallow cells).

InSTREAM provides two facilities for intermediate output. One is "probes" opened from the graphical display. These are windows that can be opened to manually observe and control the variables of individual fish, redds, and habitat cells. The second facility is a variety of optional output files that provide intermediate results for testing the model and its software.

12. Scheduling

The order of events can strongly affect the outcome of individual-based models. This section defines the schedule of events in InSTREAM. The schedule consists of an ordered list of actions, each executed once per simulation day. An action is defined by a list of objects, the methods those objects execute, and rules for the order in which the objects are processed. There are four main action groups (groups of related actions over the same list of objects): habitat, fish, redd, and observer. The full schedule is displayed at the end of this section.

12.1. Habitat Update Actions

Habitat updates are scheduled first because subsequent fish and redd actions depend on the day's habitat conditions. For each reach, time-series input data (flow, temperature, turbidity) are obtained for the current simulation date. The new flow is used to update the depth and velocity of all cells in each reach. The daily food production is calculated for each cell, and the amount consumed by fish reset to zero.

12.2. Fish Actions

Fish actions occur before redd actions because one fish action (spawning) can cause redd mortality via superimposition. This order means that new fish emerging from a redd do not execute their first fish actions until the day after their emergence. Scheduling fish spawning before redd actions also means that redds undergo all redd actions on the day they are created.

The four fish actions in the model are conducted in the following order: spawning, habitat selection, growth, and survival. Actions are carried out one fish at a time, in descending order of fish length. Each of these four actions is conducted for all fish before the next action is executed.

Spawning is the first fish action because spawning can be assumed the primary activity of a fish on the day it spawns. Spawning also affects habitat selection in two ways. First, female spawners move to a cell with spawning habitat on the day they create a redd. Second, when fish spawn, their weight and condition are substantially reduced, which affects their choice of habitat (giving higher preference to habitat providing high growth).

Habitat selection is the second fish action each day because it is the way that fish adapt to the day's new habitat conditions; habitat selection strongly affects both growth and survival. Note that both fish size and condition (which affect survival probabilities and reproductive status) affect habitat selection. Habitat selection is based on the fish's size **before** the current day's growth, because a fish's growth depends on its habitat choice.

Growth precedes survival because changes in a fish's length or condition factor affect its probability of survival.

Survival has its own subschedule because it includes evaluation of several different mortality sources. The number of fish killed by each mortality source can be affected by the order in which survival probabilities for each source are evaluated. Placing a mortality source earlier in the survival subschedule makes it slightly more likely to cause mortality (a mortality source cannot kill a given fish on a given day if a preceding mortality source kills the fish first). Therefore, widespread, less random mortality sources (e.g., high temperatures, high velocities) are scheduled first. Survival probabilities for these sources tend to be negligible (very close to 1.0) under most conditions and low when an unusual event occurs.

12.3. Redd Actions

Redd actions occur last each day because redds do not affect either habitat cells or fish (with the exception of creating new fish, as discussed above). There are three redd actions: survival, development, and emergence. These actions are applied to the existing redds in the order in which the redds were created, but this order has no effect on redds or newly emerged trout.

Redd survival is the first redd action to be executed. Survival is scheduled before emergence so that fish within redds are subject to redd mortality on the day they emerge. Otherwise, emerging fish would risk neither redd mortality nor fish mortality that day. Redd survival includes five separate egg mortality sources that follow their own subschedule. The redd mortality sources are scheduled from least random (extreme temperatures) to most random (superimposition).

Development follows survival, with emergence third. Because development precedes emergence, new fish begin to emerge on the same day a redd completes development.

12.4. Observer Actions

Observer actions collect and record data on the digital world inside InSTREAM. Because the output produced by observer actions is the only information that users have about the complex events going on inside the model, fully understanding model results requires knowing how observations are scheduled with respect to other model actions.

Observer actions are the last of the daily model actions. Therefore, the model's graphical and file outputs represent the state of the model after all the habitat, fish, and redd actions have been completed for a day. This scheduling means, for example, that the size and condition of a fish observed from the graphical user interface reflects the fish's state after it has completed its daily feeding and growth, not its state when it made its habitat selection or spawning determination.

12.5. Complete Schedule

Figure 40 displays the four main action groups and the actions within each group, in the order they are executed on each daily time step.

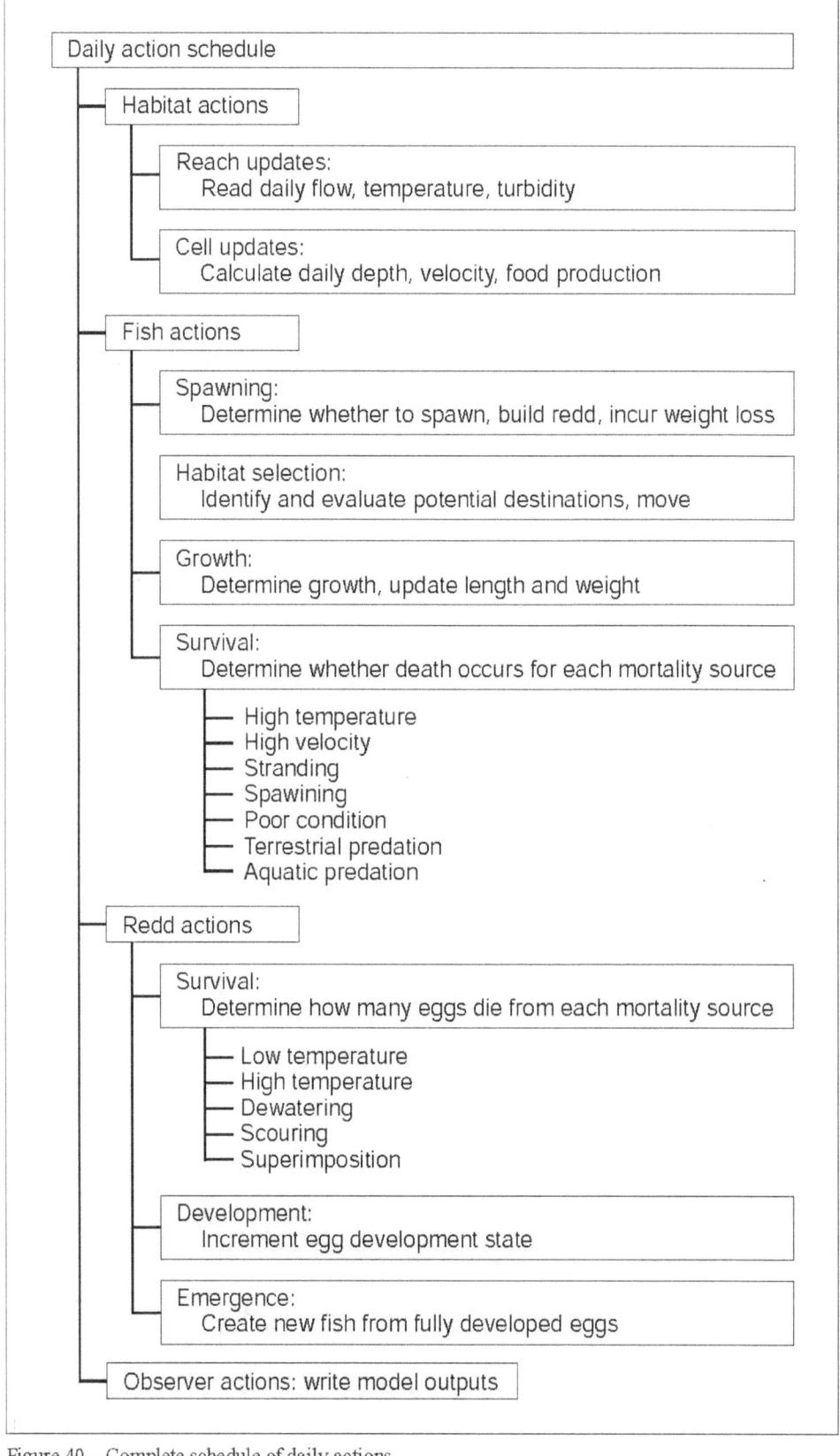

Figure 40—Complete schedule of daily actions.

Chapter 3: Model Application

13. Application of InSTREAM: Introduction and Objectives

Users of InSTREAM must understand the model's assumptions and methods (the subject of chapter 2) and know how to operate its software (the subject of chapter 4). It is also critical for users and potential users to understand how InSTREAM is best applied to particular study sites and river management (or research) problems. This understanding is important for deciding whether InSTREAM is appropriate for specific problems and for planning and conducting applications with efficiency and credibility.

This chapter has five sections that provide guidance for applying InSTREAM to specific sites and study questions. Section 14 discusses the design of studies that use InSTREAM. Study design—the general approach for applying the model to a study question—is addressed first because it affects data collection and calibration. Section 15 provides a general discussion of sensitivities and uncertainties in InSTREAM, another topic that can influence how input is assembled and how the model is calibrated. Sections 16 and 17 address two main activities in applying any environmental model: assembling the input and calibrating the model.

14. Study Design

Models like InSTREAM are applied to management or research problems by designing and conducting **simulation experiments**: controlled experiments conducted on the model, just as scientists conduct controlled studies in the field or laboratory. In many ways designing simulation experiments is similar to designing field studies—the study design must determine:

- Which inputs (data or parameters) to control (hold constant) and which to vary in what ways.
- What model output must be "observed" from the experiment.
- How many replicates are needed.
- What statistical or graphical methods will be used to interpret the data.

One key difference between simulation experiments and field studies is that simulation experiments are cheap, fast, well-controlled, and highly observable. Instead of relying on extensive statistical analysis of relatively few data (the typical situation with field studies) we can run more simulation experiments. Ideally, modeling is conducted in conjunction with field studies so simulation results can be used to design field experiments that then test and improve the model. Grimm and Railsback (2005; see their chapter 9) provided general guidance on study design and analysis of individual-based models such as InSTREAM.

Two terms (also defined in section 3) are important in discussing study design. A **scenario** is a complete set of input to InSTREAM that represents one particular set of environmental conditions and one management alternative. Effects of various environmental conditions (e.g., wet, average, or dry flow years) or management alternatives (e.g., instream flow rules A, B, or C) are typically assessed by comparing output produced by several different scenarios. **Replicates** are multiple model runs that represent the same scenario (use exactly the same input data and parameters) but use different pseudorandom number sequences to alter the outcomes of the model's stochastic processes.

Some issues to consider in designing applications of InSTREAM are:

- The model produces many kinds of output, including time series of population abundance and biomass broken out by age and species, mortality rates, habitat availability and use, and the number, development, and success of redds. Simulations of multiple species can predict the relative abundance of each species, and long-term simulations can predict the frequency of local extinction under various conditions.

- Model results are stochastic, mainly because mortality of fish and redds are stochastic. How important are the differences in model results for different scenarios, given that some differences are due to chance?

- Model assumptions, parameters, and input data are uncertain, so model predictions are uncertain. How can the model best support decisions, considering its uncertainties?

- Natural variability in physical habitat (river channel shape, etc.) and in flow and weather can have strong effects on results, sometimes stronger than those of the management alternatives the study is designed to compare.

- Although simulation experiments using InSTREAM are far faster and cheaper than field studies, there are practical limits on how many model runs can get done and how big and long the runs can be.

- The arbitrariness of conventional statistical methods for comparing scenarios (Hilborn 1997) becomes especially apparent with data generated by a simulation model. Whether the results of two scenarios are "significantly different" depends highly on how many replicates are used and how different the scenarios are; both of these factors are easily manipulated in simulation experiments.

The importance of each of these issues depends on the study site and the problem addressed. The following subsections provide general guidance developed from early applications of InSTREAM.

14.1. Outputs to Analyze

One of the first questions in designing a study using InSTREAM is which of the model's outputs to analyze. Usually it is practical to analyze how several outputs differ among scenarios, but outputs can differ in various and sometimes conflicting ways. Often, focus on only one or two key outputs provides clearer, less-confusing support for management decisions. Of course, checking other outputs (e.g., causes of mortality, redd success, frequency of "extinction" of the simulated population) may reveal factors having strong effects on the simulated trout population that are not represented well by the one or two basic outputs.

Especially for management decision support such as instream flow assessment, the study design focuses on the abundance and size of adult trout. In fact, population biomass (total weight of all fish, determined by multiplying abundance by mean weight) is a useful measure that reflects both abundance and size of individuals. Abundance and size are clear, measurable indicators of the status of fish populations; managers and the public can easily relate to these indicators, and comparable measures of real populations are often available.

Of course other study goals may require focus on other outputs. If the viability of a threatened trout population is the key management interest, then simulated population persistence may be the most important model output. Persistence can be evaluated as, for example, the number of replicate 50-year simulations in which extinction occurred, out of 50 total replicates. In an attempt to estimate effects of an introduced species, the relative abundance of that species versus a native species may be as important as absolute abundance.

14.2. General Experimental Designs

This section briefly describes two common general designs for studies using InSTREAM: scenario comparisons and sensitivity analyses. Scenario comparisons resemble the study designs commonly used in field or laboratory studies, but sensitivity analyses often can provide more information about how and why fish populations are expected to vary.

14.2.1. Scenario comparisons—

Scenario comparisons are more analogous to traditional field studies: several distinct scenarios are defined, simulated, and compared. This approach is natural when the purpose of the study is to compare and rank a few distinct management actions: assessing several alternative rules for instream flow releases, examining the effects of introducing another trout species, or deciding which of several channel restoration projects to implement.

The first step in conducting scenario comparisons is to define the scenarios by specifying what model inputs (parameter values and data such as channel shape, flow, and temperature, and, possibly, how many and which species) will vary among scenarios. For example, if the objective is to compare three sets of alternative instream flow release rules for a reservoir, then defining the scenarios could require modeling (e.g., with a reservoir water balance model) the time series of daily flows that would result from each alternative rule set. If water temperature at the study site is strongly affected by flow, defining the scenarios could also require modeling a time series of daily temperatures corresponding to each daily flow time series. In another example, Harvey and Railsback (2009) compared multiple turbidity scenarios, each differing only in the assumed relation between turbidity and streamflow. Harvey and Railsback defined the scenarios using field measurements of streamflow and turbidity. The lowest turbidity scenario represented an undisturbed watershed in which turbidity increased with flow with a low slope (e.g., 25 NTU turbidity at flow of 1.0 m^3/s), and the highest turbidity scenario represented a highly disturbed watershed with rapid increases in turbidity with flow (e.g., 190 NTU at 1.0 m^3/s).

The second step in a scenario comparison study is to decide what kind of replication to use. Normally, replication of stochastic simulation models is done simply by altering the random number generator seed, which affects the model's processes that depend on pseudorandom numbers. This standard random-number replication evaluates how much model results are affected by the processes that are assumed to be stochastic, and is the most common way to replicate InSTREAM results. But there is at least one other type of replication to consider: InSTREAM's "year shuffler" facility (described in chapter 4) makes it easy to randomly reorder the years of input data. Simulations replicated by randomizing the input data years examine how natural variability in flow, temperature, and turbidity (in addition to the model's stochastic processes) affect results.

Next, users must decide on the appropriate level of replication. Using too few replicates makes it impossible to understand how much of the variation among scenarios is due only to stochasticity, but using too many replicates can produce highly statistically significant results even though their biological significance (or the likelihood of ever measuring the effect in the field) is low. No clear guidance addresses the appropriate number of replicates. Using around 5 to 10 replicates has appeared sufficient to identify differences among scenarios that seem likely to be important for management. Five to 10 replicates are insufficient to precisely define the distribution of stochastic results for a scenario, but if two scenarios cannot be distinguished by this many replicates then their difference seems unlikely to be important compared to the effects of natural variability and model uncertainty. Besides being computationally burdensome, large numbers of replicates may encourage overestimation of differences among scenarios. Statistical differences between two scenarios can always be found with sufficient replication, whether or not the difference is biologically meaningful.

Results may be analyzed statistically or simply displayed. Statistical comparisons (typically, using two-sided t-tests or analyses of variance) can be informative, especially for audiences accustomed to statistical analysis. Results of replicated scenarios very likely meet the assumption of normality for parametric statistics, but normality of results can be tested. However, simple graphical comparison of results (e.g., via bar charts with error bars, or scatter plots showing the value of each replicate—see Magnusson 2000) are often just as informative and easier to interpret. For examples of both statistical and graphical comparison of scenario results, see Railsback and Harvey (2002) and Railsback et al. (2005).

14.2.2. Sensitivity analyses—

Sensitivity analyses are experimental designs intended to provide more and broader information about fish population responses to management actions. Instead of comparing a few discrete scenarios, a sensitivity analysis provides an "incremental" analysis of how the trout population would respond over a range of actions. For example, an analysis could look at the population's sensitivity to summer instream flow by simulating a broad range of instream flow scenarios.

The general steps in a sensitivity analysis are similar to those for a scenario comparison. First is to identify and define the scenarios to be simulated. A large number of scenarios are typically produced by varying only one or two inputs. An incremental analysis of instream flows would include perhaps 10 to 20 flow scenarios from very low to very high flow. As another example, a sensitivity analysis of the effects of weekend pulse flow releases for whitewater recreation could include scenarios ranging from 0 to 20 weekends of whitewater release per year. Multivariate sensitivity analyses are also possible, for example, by defining 50 scenarios that include 10 levels of minimum instream flow and 5 levels of whitewater release. It is useful to extend the simulated scenarios beyond the range considered feasible for management, so trends are understood even at their extremes.

The second step is to execute the scenarios. Typically only one replicate is executed per scenario because it is not important to understand how much variability there is in the results for each scenario: examining many scenarios lets us examine both the model's response to the input being varied and the noise in this response.

Using only one replicate of many scenarios does not mean that stochasticity in results is ignored in analyzing results. Analysis of results in a sensitivity analysis can use graphical and statistical techniques, but the techniques are different than for scenario comparisons. The first analysis step should always be to plot how the trout population output of InSTREAM varied over the range of scenarios. Did the population increase or decrease consistently over the range? In some cases there may be little trend detectable, indicating that the "signal" from the variable that changed among scenarios is small compared to the stochastic "noise" in the results. If a multivariate sensitivity analysis was performed, then contour plots can be useful to examine responses.

14.3. Variation in Inputs: Realistic Versus Unrealistic Scenarios

Another basic study design decision is how much realistic variation to include in the input scenarios. One advantage of IBMs such as InSTREAM is that they can naturally predict the effects of realistic, day-to-day variation in the driving inputs: flow, temperature, and turbidity. These variables typically vary from day to day as well as seasonally, even downstream of major reservoirs. However, including more variation in a simulation experiment can make it more difficult to thoroughly understand its results. Therefore, designing studies requires choosing between using scenarios that include realistic levels of variation in the time-series inputs, or using unrealistic scenarios in which some or all of the variation in some of the inputs is suppressed to make analysis easier.

In general, if the purpose of a study is assessment of management actions, there is less interest in understanding the details of how results arose and more interest in just predicting how trout populations respond to the actions. In this case, scenarios with realistic levels of variation may be most appropriate because they compare management actions in a simulated context most closely resembling the real situation.

If, however, the purpose of a study is more focused on understanding the processes by which river management affects trout populations, then experiments using scenarios with unrealistically reduced variation of some variables may be most appropriate. For example, Railsback and Harvey (2002) conducted a simulation experiment with InSTREAM to examine seasonal variation in trout habitat preferences. To focus the experiment only on the effects of two seasonal variables—temperature and day length—they unrealistically assumed other variables such as flow and trout density were constant among scenarios.

14.4. Robustness of Study Results

A key issue in any modeling study is understanding the sensitivity of results to modeling assumptions and inputs: How different would the results be if the model used a different equation, or was calibrated differently, or used different input? Traditionally these questions have been thought of in terms of model sensitivity and uncertainty, but it is more productive to think of them in terms of robustness: How robust are the conclusions drawn from a modeling study (see section 9.7 of Grimm and Railsback 2005)? This robustness question is especially important (and potentially controversial) for complex models such as InSTREAM, but it cannot be answered without first answering several other questions:

- Robustness of what? Are we interested in the robustness of the primary predictions of InSTREAM—the simulated trout population abundance, production, etc., or of secondary predictions such as the predicted differences among scenarios or the predicted sensitivity of population status to variables such as instream flow or temperature?
- Robustness to what? Are we interested in the robustness of results to equations and assumptions (e.g., which processes are included versus ignored in InSTREAM), or to parameter values, or to input data?
- In what context? The robustness of results from InSTREAM undoubtedly varies with the conditions simulated. For example, results may be insensitive to equations and parameters for temperature mortality when InSTREAM is applied to a site where temperatures never exceed 15 °C, but very sensitive to these assumptions at sites with higher temperatures.

Section 15 discusses one class of robustness: the sensitivity of primary results of InSTREAM to parameter values. Unfortunately, the results of the analyses in section 15 cannot be assumed applicable to all sites, and repeating the analyses at new study sites would be a substantial burden.

Evaluating the robustness of study results to the basic modeling assumptions and equations of InSTREAM will usually be beyond the scope and capability of routine applications. Making such an evaluation would require identifying reasonable alternative assumptions, implementing them in the software, testing them, and then analyzing the results of the alternative assumptions. This kind of analysis certainly represents interesting and valuable research but is impractical for most studies.

What kinds of robustness analysis are practical and valuable for routine applications of InSTREAM? For study designs involving comparison of several alternative scenarios or analyzing sensitivity of predicted trout populations to a small number of variables, it should be practical to conduct additional model runs and examine how robust the most important results are to a few key inputs. Most important is determining whether and how the final ranking of management alternative scenarios changes as key inputs are varied. (Drechsler et al. 2003 provided an example for a different kind of example.) These key inputs could include:

- Values for a small number of the parameters identified in section 15 as having the strongest effects on the primary predictions of InSTREAM.
- Values of any additional parameters expected to be particularly important for the specific study, for example, parameters controlling sources of high mortality among fish and eggs.
- Hydrologic and weather conditions: years with high versus low base flows, more versus fewer extreme flow events, warm versus cool temperatures, etc. Rates of mortality among simulated fish and eggs can again be used to identify important inputs to evaluate.
- The sequence in which different year types occur in the input (see discussion of the "year shuffler" in section 9).

14.5. Conclusions and Summary Guidance

Outlining an appropriate study design should be one of the first steps in applying InSTREAM to either river management or research studies. Not only does the study design influence the kinds of inputs and analyses needed, but the process of developing the study design helps clarify how the modeling work can support the management decisions or research objectives.

The most important point in designing studies that use InSTREAM is the concept of simulation experiments: given appropriate input and calibration, the model can be considered a laboratory in which we conduct controlled experiments.

Many applications of InSTREAM are expected to have the objective of comparing several distinct management alternatives, for example alternative instream flow policies that determine (along with reservoir characteristics and inflows) daily flow releases. The most appropriate study design for such situations is usually the scenario comparison approach: specifying sets of model input to represent each of the alternatives and using replicate simulations to examine the degree of predicted difference in trout populations among the alternatives. Model results can be used to rank the alternatives by their predicted benefits to trout (e.g., simulated average annual population biomass). The robustness of the analysis can be examined by determining if and how the rankings change when simulations use different values of key parameters, or different weather and hydrologic conditions, or alternative values for other particularly uncertain inputs.

For many studies, it will also be useful to conduct sensitivity (or "incremental") analyses of how trout populations are predicted to vary over a broad range of inputs such as flow or temperature. Sensitivity analyses use many scenarios with only one or two inputs varied gradually over wide ranges. These analyses are more useful for developing understanding of how management variables affect trout and for finding good management policies. Results can be analyzed by developing (graphical and perhaps also statistical) relationships between measures of the trout population and values of the inputs. Robustness of results can be examined by determining if and how these relationships change when different values of particularly uncertain and important inputs are used. Especially for sensitivity analyses, users should not feel constrained to using "realistic" input sets that include all possible kinds of variability. Instead, simulation experiments that unrealistically limit variability in some inputs can be helpful for understanding some problems.

15. Sensitivities, Uncertainties, and Robustness of InSTREAM

This section reports studies conducted at Humboldt State University to analyze sensitivities and uncertainties of InSTREAM, using techniques modified from conventional model analysis techniques (e.g., Saltelli et al. 2004). The analyses addressed sensitivity of the model's primary predictions (simulated trout population abundance and biomass) to parameter values, including interactions among key parameters. But the studies also addressed a question more relevant to the use of inSTREAM for management decisionmaking: How robust are management conclusions drawn from InSTREAM to the values of key parameters? Can InSTREAM produce robust management results even with its parameter uncertainty? Other analyses looked at how sensitive InSTREAM results are to physical characteristics of study sites.

Overall, the results of the uncertainty and robustness analyses are encouraging. We found InSTREAM generally sensitive to the physical habitat variables it was designed to predict the effects of; a lack of such sensitivity would mean InSTREAM is not useful for its intended objectives. Primary predictions were highly sensitive to only a few parameters, none of which were unexpected.

Perhaps the most important conclusions drawn from the sensitivity and uncertainty studies are: (1) InSTREAM does not appear vulnerable to runaway "error propagation," or extreme and unexpected sensitivities. Users of InSTREAM can be confident that small changes in parameter values or other inputs will not produce unexpectedly extreme changes in results. (2) Management results from inSTREAM—how it ranks the relative fish benefits of stream management alternatives—appear quite robust to parameter uncertainty.

15.1. General Sensitivity Considerations

Conventional parameter sensitivity and uncertainty analysis methods (e.g., Saltelli et al. 2004) are not directly applicable to IBMs such as InSTREAM for several reasons:

- The high number of parameters in InSTREAM, combined with the computation time for each model run, precludes a comprehensive examination of interactions among all combinations of parameters.
- IBMs have many kinds of outputs, so a comprehensive examination of how parameters affect all model results is impractical. Instead, it is only practical to examine sensitivity of selected important outputs.
- Because results are stochastic, parameter sensitivity must be separated from random noise.
- The strength of a parameter's effect on results can vary with the site and conditions being modeled. A sensitivity analysis for one site can produce quite different results than the same analysis conducted for a different study site. For example, parameters controlling effects of high temperature on survival or metabolism can have little effect at sites with benign temperature regimes, but great effects at sites with high temperatures.
- Many parameters affect model results nonlinearly, so their influence can vary sharply with their value. At least one parameter (*fishFitnessHorizon*) has non-monotonic effects: as its value increases, predicted trout populations first increase, then peak and decrease.
- The model's usefulness for management decisionmaking may depend less on its primary quantitative predictions (e.g., trout population biomass) than on the ranking of several management scenarios. The robustness of these rankings to parameter values can be more important than the sensitivity of primary predictions.

Therefore, we developed new strategies for analysis of large management-oriented IBMs (Butcher and Parrish 2006; Cunningham 2007; Railsback et al., n.d.) that are designed to develop as much important information as possible within the computational constraints. The following sections are based on these new strategies.

15.2. Sensitivity of Primary Predictions to Parameter Uncertainty

This section provides information on the sensitivity of primary predictions of InSTREAM (specifically, simulated total biomass of the adult trout population) to parameter values. This information is especially useful for identifying parameters that (1) are best to use for calibration (discussed in section 17.1) and (2) deserve special attention when applying InSTREAM.

Cunningham (2007) conducted sensitivity analyses using conventional single-parameter perturbation methods. For each parameter, a range of feasible values was identified, and the model run for seven values over that range. All other parameters were held constant at their standard value. To analyze results, the parameter values were all scaled over a range of 0 to 1: the lowest parameter value has a scaled value of 0.0, the standard ("best") value has a scaled value of 0.5, and the highest parameter value was scaled to 1.0. Then regression was used to determine the slope of model output with respect to the scaled parameter values. The absolute value of this slope was used as an index of model sensitivity to the parameter. (For one parameter that yielded a non-monotonic model response, sensitivity was evaluated using the slope of the steepest part of the hump.)

The sensitivity analysis examined mean total biomass of adult (age 2 and older) trout over the last 11 years of a 14-year period. One output of adult biomass was obtained for each year (near October 1) and used to compute a mean for the entire model run.

The analysis used the lower study reach at Little Jones Creek (Smith River watershed, Del Norte County, California; Harvey 1998). This reach has moderate temperatures, low turbidity, a maximum depth of approximately 1.5 m, and velocities generally below 1.5 m/s at typical flows. Hence, this analysis cannot be assumed to represent how sensitivite InSTREAM is to habitat-related parameters at sites with more extreme habitat conditions.

The parameters to which InSTREAM was determined to be most sensitive in this analysis are identified in table 23. Table 23 also explains why the model is sensitive to the parameters, parameter uncertainty, and therefore how the parameter should be treated in parameterizing and calibrating InSTREAM. A parameter may not be of special concern if its value is well known, even if the model is highly sensitive to it. On the other hand, the parameters in table 23 that do not have well-known values, or that represent inherently variable and uncertain processes, deserve special attention. The complete results (table 24) show that the vast majority of parameters had little effect under the conditions simulated by Cunningham (2007)—but most of these parameters could have strong effects under some conditions.

Table 23—Parameters to which inSTREAM was found most sensitive in the analysis of Cunningham (2007), in order of decreasing sensitivity

Parameter	Sensitivity considerations
mortFishTerrPredD9	If this parameter is set to a low depth (or close to mortFishTerrPredD1), depth offers very high protection and terrestrial predation becomes negligible in many cells. The parameter is expected to have much less effect when set to higher values.
	Values are not well known and can vary with predator types: birds may be less effective on fish at depth than are otters, for example. Normally, the value should be set so no habitat is routinely immune to terrestrial predation. Values should be selected for each study site, before calibration.
habPreyEnergyDensity	Trout energy intake increases linearly with this parameter, and is not limited by the maximum daily intake (cMax).
	Energy density of invertebrate prey can vary seasonally as prey types change, but the range of reasonable values is well-known. Values should be selected for a study site before calibration.
fishRespParamB fishRespParamA fishRespParamC	Respiration parameters strongly affect energy costs and growth.
	Values are relatively well-known from laboratory studies, and typically should not be changed.
fishWeightParamA fishWeightParamB	Seemingly small changes can greatly affect the growth in length that results from growth in weight.
	Values can vary slightly among sites; using values from field data or literature will prevent significant error.
habDriftConc	Energy intake increases linearly with this parameter, until intake is limited by cMax.
	Values are site-specific, rarely well-known, and this parameter also represents a variety of simplifications and uncertainties in food availability, so there is no guarantee that measured values will produce useful model results. This parameter is best evaluated via calibration to observed growth or size.
fishMaxSwimParamA fishMaxSwimParamD	These affect both food intake (how capture success varies with velocity) and velocity mortality. Consequently, they strongly affect how many cells offer positive growth and high survival.
	Values are from laboratory studies, but are moderately uncertain owing to variability among individuals and measurement difficulties. Values should typically not be changed.
fishDetectDistParamB	For small trout, small changes can produce large changes in drift food intake, which increases with the square of the parameter.
	Values are based on laboratory data, but detection distance depends on factors (especially prey type and size) not considered in inSTREAM. The value was estimated partly by calibrating its effect on relative growth of small vs. large trout. Normally, the value should not be changed.
habDriftRegenDist	This parameter controls total food availability in a cell.
	Values are highly uncertain, as this single parameter represents a highly variable process that is difficult to measure. Values are best obtained via calibration of fish density in high-quality cells.
mortFishTerrPredMin	Terrestrial predation is normally the most important mortality source for trout more than a few centimeters in length.
	Values are highly uncertain and variable, so are best estimated via calibration to observed survival and abundance.

Note: Chapter 2 explains the parameters and their values.

Table 24—Complete sensitivity results for Little Jones Creek (Cunningham 2007), as a percentage of the maximum sensitivity

Parameter	Sensitivity	Parameter	Sensitivity	Parameter	Sensitivity
fishSpawnMinLength	0	reddDevelParamC	2	mortFishAqPredMin	6
mortFishAqPredF1	0	mortFishAqPredD9	2	fishCaptureParam9	6
mortFishAqPredF9	0	mortFishTerrPredV1	2	habShelterSpeedFrac	7
mortFishAqPredT1	0	mortReddLoTT9	2	habSearchProd	8
mortFishAqPredT9	0	mortReddHiTT1	2	mortFishAqPredL9	9
mortFishAqPredU1	0	mortFishTerrPredL1	2	fishRespParamD	10
mortReddDewaterSurv	0	mortFishAqPredP9	2	mortFishTerrPredH9	10
mortFishTerrPredT1	0	fishSpawnEggViability	2	fishMoveDistParamA	11
fishCmaxParamA	0	mortFishStrandD9	2	mortFishConditionK1	12
fishFitnessHorizon	0	mortReddScourDepth	2	fishMoveDistParamB	13
reddNewLengthStdDev	0	mortFishAqPredD1	2	fishMaxSwimParamB	13
fishTurbidMin	0	fishMinFeedTemp	3	fishMaxSwimParamE	14
mortReddHiTT9	0	mortFishTerrPredF9	3	mortFishConditionK9	14
mortFishAqPredU9	1	fishSpawnWtLossFraction	3	mortFishTerrPredH1	15
fishCmaxParamB	1	mortFishStrandD1	3	fishCaptureParam1	15
fishFecundParamB	1	mortFishTerrPredD1	3	fishMaxSwimParamC	16
fishSpawnMaxFlowChange	1	reddDevelParamB	3	fishEnergyDensity	16
habShearParamB	1	reddNewLengthMean	3	mortFishTerrPredMin	16
mortFishAqPredP1	1	fishSearchArea	3	fishRespParamC	22
mortFishHiTT9	1	fishSpawnProb	3	habDriftRegenDist	23
mortFishVelocityV9	1	fishDetectDistParamA	3	fishRespParamA	24
mortFishTerrPredT9	1	mortFishHiTT1	4	fishMaxSwimParamD	26
mortFishTerrPredF1	2	mortFishTerrPredV9	4	fishDetectDistParamB	29
mortReddLoTT1	2	fishSpawnMaxTemp	4	fishMaxSwimParamA	29
fishSpawnMinCond	2	mortFishAqPredL1	4	habDriftConc	30
habShearParamA	2	fishSpawnMinTemp	5	fishWeightParamA	30
habMaxSpawnFlow	2	mortFishVelocityV1	5	fishRespParamB	32
mortFishTerrPredL9	2	reddSize	6	habPreyEnergyDensity	55
reddDevelParamA	2	fishPiscivoryLength	6	mortFishTerrPredD9	100
fishFecundParamA	2	fishTurbidExp	6		

Note: Sensitivity values and ranking are expected to differ substantially among sites.

15.3. Sensitivity of Primary Predictions to Initial Populations

Sensitivity of predicted trout abundance to the number of trout at the start of a simulation was investigated by varying the initial abundance in nine otherwise identical scenarios. This experiment used input from the Little Jones Creek lower mainstem study reach. Initial abundances (for October 1) ranged from 10 to 400 percent of the "real" values (obtained from field censuses) of 186 age 0, 28 age 1, 9 age 2, and 7 age 3+ trout. Simulations were then run for 11 years, with five replicates of each scenario.

This sensitivity experiment showed that effects of initial abundance were unimportant by the third simulation year, except when initial abundance was less than half the correct value (fig. 41). The effect of even extremely high initial

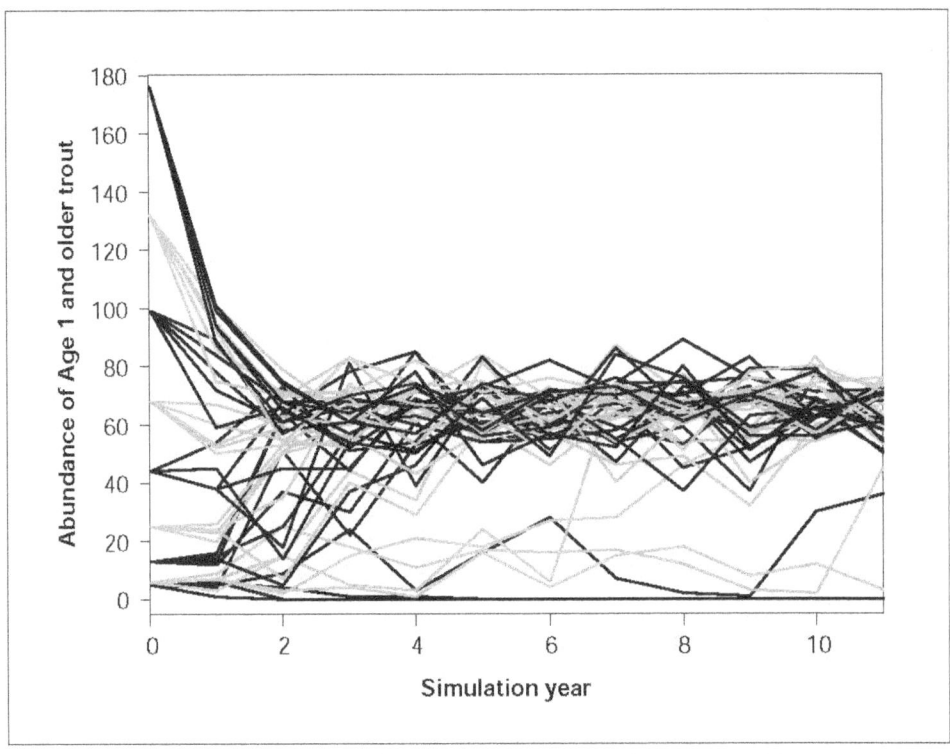

Figure 41—Time series of simulated abundance of age 1 and older trout, for nine initial abundance scenarios. Five replicates of each scenario are shown. The "real" initial abundance is 44. Simulations used input for the Little Jones Creek lower study site, where total abundance and spawning success are relatively low.

abundance appears to be gone within 3 years. On the other hand, model runs with initial abundance much less than half the "real" value can take many years to build up to normal levels. This sensitivity to low initial abundance may not occur at other sites where total abundance is higher and spawning success more reliable. Trout populations can recover rapidly given high spawning success.

This analysis suggests that population estimates used to initialize the model should tend to overestimate instead of underestimate abundance.

15.4. Sensitivity of Primary Predictions to Habitat Input

This section examines the sensitivity of predicted trout population biomass to site-specific habitat input: the size and spatial arrangement of habitat cells, and the input describing hiding cover, velocity shelter, and spawning gravel in cells. Sensitivity to these inputs is not a disadvantage of InSTREAM; in fact, it needs to be sensitive to such inputs to be useful for its purpose of predicting how trout populations depend on physical habitat and river management. This discussion summarizes simulation experiments conducted by Butcher and Parrish (2006), who developed a process for synthesizing habitat input sets for InSTREAM. Under this process, multiple sets of habitat input can be stochastically generated from the same general site characteristics. These site characteristics include the relative frequency of habitat types (pools, riffles, and runs), characteristic channel cross sections for each habitat type, the density of hiding cover (number of cells containing hiding cover), and the distribution (beta distribution parameters) of cell velocity shelter and spawning gravel variables.

The experiments described here used habitat characteristics of a study reach on Bull Creek, a tributary to the South Fork Eel River, Humboldt County, California. This reach is a third-order stream with a moderate (1 percent) gradient and drainage area of 2600 ha. Simulations covered 14 years; output analyzed was the mean biomass for adult trout (age 2 and higher) on October 15 over the last 11 years of the simulations.

15.4.1. Study site—

One of the key issues in using habitat-based models like InSTREAM is the influence of study reach selection: Does the specific placement of study reaches matter? This experiment addresses this issue by exploring how sensitive InSTREAM's predictions are to the exact sequence of habitat units when the statistical distributions for relative frequency and length of habitat types are held constant. This is comparable to testing how model results vary among different reaches chosen from the same stream.

Butcher and Parrish (2006) represented habitat units as three successive transects of the same habitat type, pool, riffle, or flatwater (run). Eight habitat units were represented in a 200-m reach, with the sequence of units drawn randomly from a Markov model parameterized with the relative frequency of the three habitat types observed in real streams. Habitat unit lengths were calculated from relationships between unit lengths and stream gradient from real streams, and the last habitat unit was truncated to maintain a reach length of 200 m. The shape of each of the three transects in each habitat unit was drawn randomly from a library of cross-sectional channel shapes; these shapes retained the characteristic width-depth ratios of each habitat type observed in field data. Cell habitat variables representing hiding and feeding cover and spawning gravel were drawn randomly using methods described below (sections 15.4.2 through 15.4.6).

This process was repeated to generate five stream reaches, each a different stochastic realization of the same statistical properties. This process does not guarantee that the five realizations have the same area of each habitat type. Five replicate simulations were conducted for each of the five reaches.

Results of this experiment indicate that the location of a study reach can have significant effects for the size of reach simulated in this experiment (fig. 42). For example, the mean trout biomass for realizations 1 and 5 are outside the 95-percent confidence interval for realization 3. Using longer reaches will tend to reduce variability among reaches by reducing the differences in the availability of different habitat types.

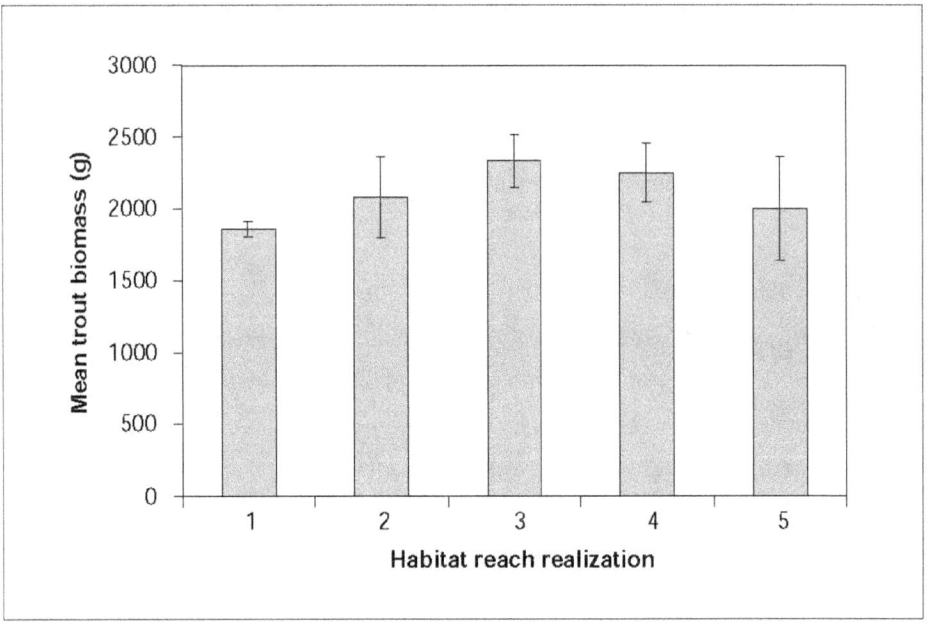

Figure 42—Trout biomass predicted for five synthesized stream reaches with the same probability distribution of habitat types. Values are mean of five replicates; error bars are 95-percent confidence intervals (2 × standard deviation).

15.4.2. Transect density—

One of the most contentious issues in stream habitat modeling (especially in the use of PHABSIM; e.g., Williams 1996) is the influence of the number of transects used to represent a study site on model results. For InSTREAM, relatively clear criteria address transect placement (section 16.3), but uncertainty remains about the number of transects needed to represent individual habitat units. Butcher and Parrish (2006) explored this question in their synthesized stream reach. They created one stream reach using the methods described in section 15.4.1, with three "true" transects used to define each habitat unit. Then they inserted new transects at regular intervals between the original ones, linearly interpolating the shape of the new transects from the shapes of the "true" transects upstream and downstream.

A series of eight new stream reaches was generated in this way with transects spaced 5, 10, 15, 20, 25, 30, 35, and 40 m apart. Cell lengths were set to the distance between transects. Then trout populations were simulated, with three replicate runs for each of the eight reaches.

Results of this experiment (fig. 43) indicate that transect density may affect predicted trout biomass. The results suggest nonlinear effects, with highest biomass at intermediate transect spacing. This effect in the range of 5 to 15 m transect spacing may be related to cell size, discussed in the next section. The decrease in biomass with long distances between transects may reflect that as the depiction of habitat is coarsened by using fewer transects, fewer cells of near-optimal habitat are likely to be available.

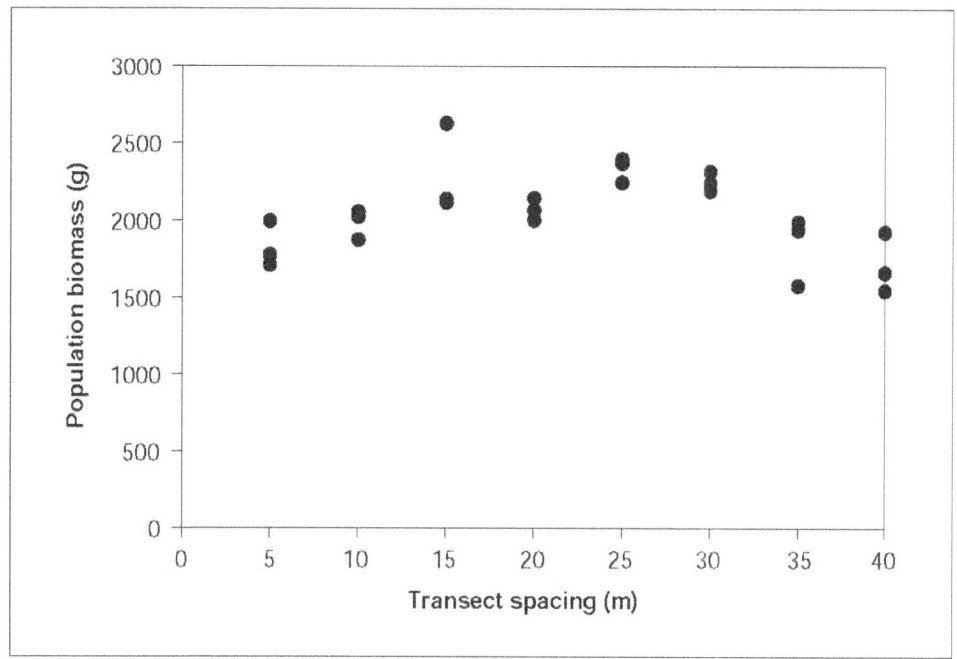

Figure 43—Sensitivity of predicted trout biomass to transect density. Each point is the mean adult trout biomass for one model run.

15.4.3. Cell size—

The assumption of InSTREAM that trout compete in a size-based hierarchy for the food in each cell raises the possibility that cell size could affect results. To illustrate: a large cell might have just enough food for three adult trout; but if it were divided into two smaller cells, each of those cells would have enough food for only one adult trout, with food left over for smaller fish. Using the smaller cell size in this case would favor fewer adults and more juveniles. However, this potential artifact is not expected to be important if cells are large enough that their food availability is large compared to the consumption of individual trout, or if small and large trout use different habitat instead of attempting to share common cells. To explore the potential effect of cell size, Butcher and Parrish (2006) created an artificial stream reach with uniform shape in the upstream-downstream dimension. They then varied the number of transects in this reach, changing only the cell length. As in the transect density experiment, cell lengths were varied from 5 to 40 m, with three replicate simulations for each cell length.

The results (fig. 44) indicate that cell size may have an effect on predicted trout biomass at small cell sizes, but no effect at moderate to large cell sizes. Although analysis of variance did not detect an overall effect of cell length on trout biomass ($p = 0.28$), the apparent increase in biomass as cell length increases from 5 to 15 m may reflect a real difference in model results.

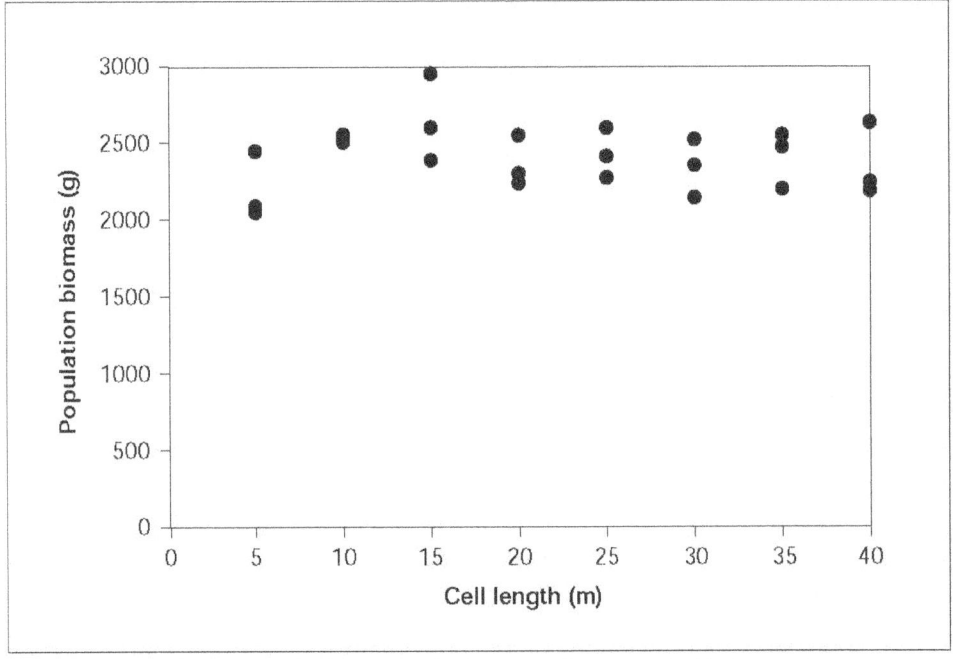

Figure 44—Sensitivity of predicted trout populations to cell length, with all other habitat characteristics held constant. Each point is the mean adult trout biomass for one model run.

15.4.4. Hiding cover—

Butcher and Parrish (2006) found that predicted trout biomass increased significantly ($r^2 = 0.92$, $p < 0.001$) and sharply as average distance to hiding cover (*cellDistToHide*) decreased. Biomass doubled, for example, when average *cellDistToHide* was decreased from 0.8 to 0.6 m. However, this experiment examined only very low values of *cellDistToHide* (less than 1 m); predicted trout populations are expected to be much less sensitive to this input at higher values because its effect on predation risk decreases as *cellDistToHide* increases.

15.4.5. Velocity shelter—

This experiment varied the values of the cell habitat variable (*fracShelter*) that represents the fraction of the cell providing velocity shelter for feeding. Butcher and Parrish (2006) drew each cell's value of *fracShelter* from a beta distribution skewed toward low values; values of zero are common in this distribution, as in typical field data. The mean of the beta distribution was varied from very low (0.001) to high (0.5) values. The predicted trout biomass increased significantly ($r^2 = 0.84$, $p = 0.01$) and sharply with the reach-averaged value of *fracShelter*, but only at values less than about 0.1. This result makes sense, as the use of velocity shelter increases trout growth, but even small values of *fracShelter* are likely to provide velocity shelter for all the trout in a cell (chapter 2).

This experiment shows the importance of accurately estimating whether each cell has some velocity shelter, whereas the exact amount of shelter is probably much less important.

15.4.6. Spawning gravel—

The experiment varied reach-average spawning gravel availability over a wide range. Spawning gravel availability did not affect trout population biomass. This result is not a surprise considering the effects of spawning gravel represented by InSTREAM (chapter 2): gravel availability can affect where a redd is placed but does not affect either the probability of spawning or the survival of trout eggs within redds.

15.5. Robustness of Management Decisions to Parameter Uncertainty

Cunningham (2007) also examined the robustness of InSTREAM–based management decisions to parameter uncertainty, using methods adapted from Drechsler (2000; see also Drechsler et al. 2003). Determining how sensitive the model's primary results are to parameter uncertainty does not directly address the more important question of how parameter uncertainty affects management conclusions drawn from the model. Management applications of a model usually involve ranking several alternative management actions by their predicted effects. This ranking of alternatives may be much less sensitive to parameter uncertainty than specific model predictions. This analysis used InSTREAM to rank four hypothetical management actions by the trout population biomass predicted for each, then examined how the ranking changed with perturbation of parameter values. This discussion uses the parameter perturbations and model results produced by Cunningham 2007 but offers a separate analysis.

This analysis included seven parameters that strongly affect model results (section 15.2): *habDriftConc, habDriftRegenDist, fishDetectDistParamB, fishMaxSwimParamA, fishRespParamA, fishRespParamB,* and *mortFishTerrPredD9*. Each parameter was varied over a range of feasible values, using Latin hypercube sampling to ensure that three (low, central, and high) regions of the parameter space were included with equal frequency. Each management action was simulated with 45 sets of parameter values (referred to as "parameter perturbations").

Table 25—Management scenarios for analysis of management decision sensitivity

Management scenario	Flow alteration	Turbidity alteration
1. Hydropower with low instream flow and unmitigated timber harvest	Minimum flow release of 0.3 m^3/s; maximum diversion of 4.0 m^3/s; no diversion when natural flow is less than 0.5 m^3/s	160 percent of baseline
2. Hydropower with high instream flow	Minimum flow of 0.5 m^3/s; maximum diversion of 4.0 m^3/s; no diversion when natural flow is below 0.7 m^3/s	160 percent of baseline
3. Hydropower with low instream flow, mitigated timber harvest	Same as scenario 2	120 percent of baseline
4. Baseline (no development)	None: natural flows	None: natural turbidities

The four management actions are hypothetical scenarios for hydropower development and timber management in the Little Jones Creek watershed. These scenarios differ in how flow and turbidity would be altered (table 25). Predicted biomass of trout age 2 and older for the last 11 years of 14-year simulations was used to rank the alternatives.

The predicted trout biomass varied widely over the 45 parameter perturbations, with the standard deviation among combinations exceeding the mean biomass (fig. 45). However, the ranking of the four scenarios was relatively robust to parameter uncertainty (fig. 46). The best management alternative was scenario 4 for 80 percent of the parameter perturbations, and scenario 1 was ranked worst in 82 percent of perturbations. The second- and third-ranked alternatives were less clearly distinguished, not surprising because of the small difference in mean biomass between them (fig. 45). Still, these two scenarios were ranked correctly for 60 percent of parameter perturbations.

Overall, the correct ranking was obtained from 56 percent of the parameter perturbations, despite the closeness of scenarios 2 and 3. The more important results—identifying scenario 1 as the worst alternative and scenario 4 as the best— were obtained from 73 percent of the parameter perturbations.

This analysis indicates that management conclusions drawn from InSTREAM can be relatively robust even when parameter uncertainty is assumed to produce high uncertainty in the model's primary predictions of trout biomass.

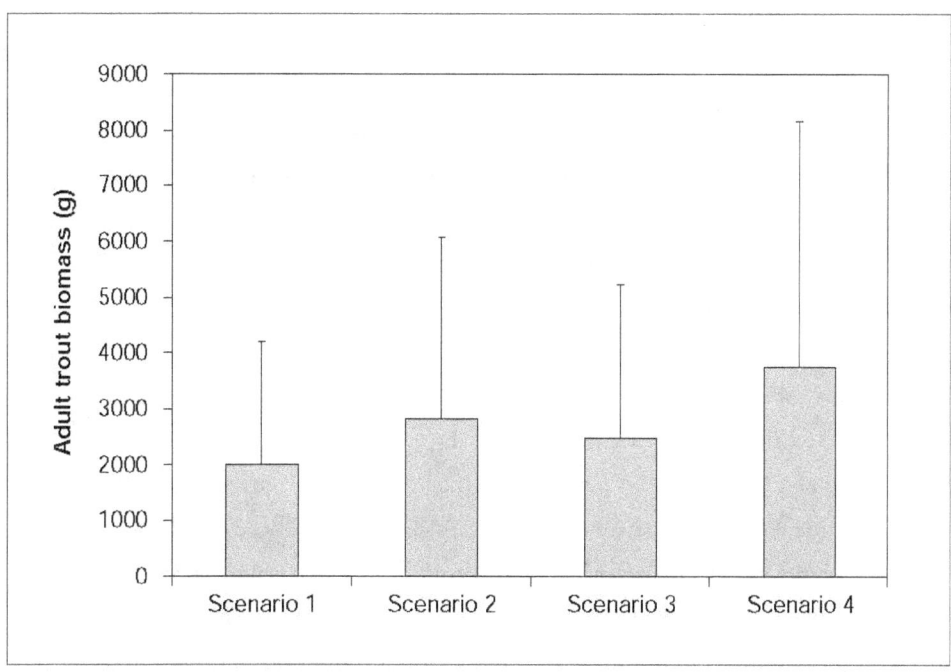

Figure 45—Mean adult trout biomass over all parameter combinations, for the four management scenarios. Error bars are one standard deviation above the mean.

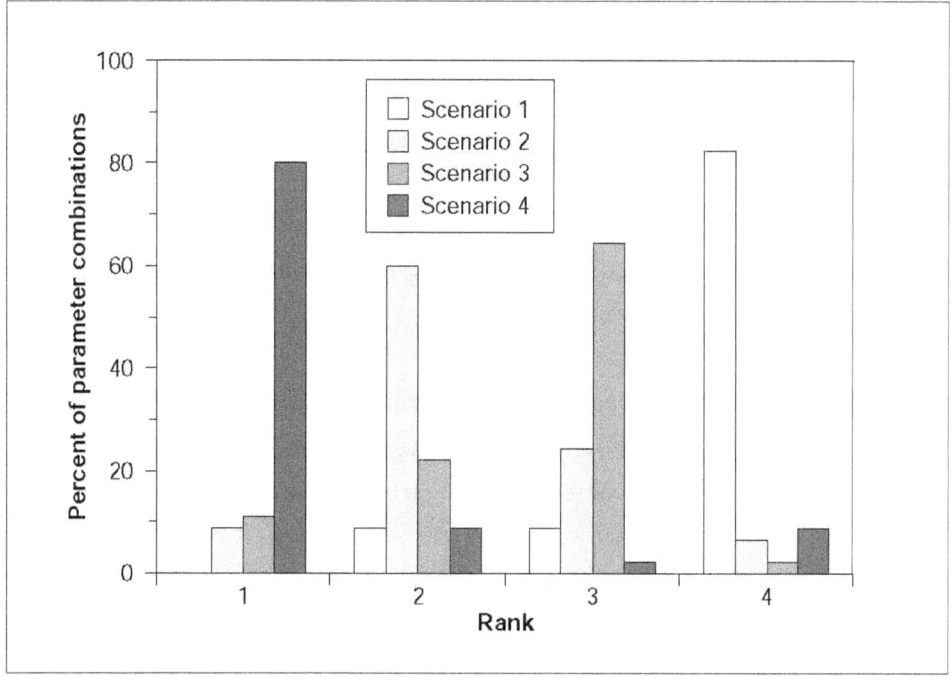

Figure 46—Ranking of management scenarios in parameter uncertainty experiments. Bars represent how frequently the rank was held by a particular management scenario. For example, scenario 4 was ranked first (highest trout biomass) for 80 percent of parameter combinations, and scenario 1 was ranked last for 82 percent of parameter combinations.

16. Assembling Input for InSTREAM

This section provides guidance on assembling the field data and parameter values required to apply InSTREAM to a new study site. Major steps in this process include selecting a study reach, collecting the time series inputs (flow, temperature, and turbidity), defining the habitat cells, measuring and modeling cell hydraulics, evaluating reach-level parameters, determining which trout species to simulate, estimating initial populations, and evaluating site-specific trout parameters. Although this list may sound daunting, experience with small to midsized streams that can be waded at most flows indicates that the initial reach setup can be accomplished in 2 to 3 days, with additional field days to take hydraulic measurements at other flows and to service the instruments that collect time series data.

The kinds of input required for InSTREAM superficially resemble those required for studies that use PHABSIM (Bovee et al. 1998), and PHABSIM hydraulic models are used to prepare input for InSTREAM. However, many important differences in data collection distinguish the two approaches. These differences result from fundamental differences in the two kinds of model (PHABSIM attempts only to model habitat quality) and conceptual flaws in how PHABSIM is typically used (Railsback 1999a). Users of InSTREAM should not let any experience with or guidance for PHABSIM override the approach described here.

This section addresses how to collect input, and the software user guide (chapter 4) provides details on formatting data and parameter files.

16.1. Flow, Temperature, and Turbidity Input

Time series of daily values for flow, temperature, and turbidity are the main environmental "drivers" of InSTREAM. We discuss these inputs first because often they can be collected even before selection of specific study reaches.

InSTREAM simply reads in daily values for the time series inputs. Most of this section discusses collecting field data, but other methods can be used to develop input: using flows estimated by a reservoir or hydrologic model, or simply scaling them from a nearby gage, or regression modeling of water temperature from weather data and flow. Summarizing the extensive literature on estimating or modeling flow, water temperature, and turbidity is beyond the scope of this report.

Instruments for monitoring water temperature and stage (water surface elevation) are now quite inexpensive. However, these instruments are often not highly reliable owing to theft and vandalism, loss during high flows, and instrument or battery failure. Using several at different locations near the study site is usually a wise investment.

One difference between PHABSIM and InSTREAM is that the transects used in InSTREAM cannot be assumed to provide useful flow measurements. Instead, a separate flow monitoring station should be set up within or near the reach at a location chosen for accurate stream gaging. Maintaining the station requires installing and maintaining a stage monitor and developing a stage-discharge relation by measuring flow over a range of stages. In some settings, an alternative to installing a stage recorder is to take sufficient measurements of flow over a wide enough range to build a reliable relationship between flow at the study site and flow at a nearby permanent gaging station.

The availability of inexpensive and accurate temperature loggers makes temperature monitoring relatively easy and reliable. One consideration for temperature input is whether it will be desirable to model how water temperature varies with flow—especially likely for studies designed to predict the effect of alternative instream flows. If temperature modeling is warranted, then additional data such as air temperature and windspeed are likely to be needed.

Turbidity is generally not as easily monitored as temperature or flow, and the temptation to ignore its effects can be strong. However, turbidity has strong and well-defined effects on trout feeding and predation risk (section 6), which means it can strongly affect how trout populations respond to changes in flow, temperature, and other habitat characteristics. In the absence of consistently low turbidity, it is important to develop reasonably accurate input data.

Field turbidity monitors require maintenance and calibration. Turbidity is loosely defined as light-scattering ability, and different instrument designs can produce different results. Information on turbidity monitoring is available from the U.S. Geological Survey's "National Field Manual for the Collection of Water Quality Data" chapter A6; and from the turbidity research Web site maintained by the U.S. Forest Service's Redwood Sciences Laboratory (http://www.fs.fed.us/psw/topics/water/tts/).

Where turbidity results primarily from suspended sediment, turbidity can be modeled as an increasing function of flow. In some systems, however, phytoplankton can elevate turbidity at low flows. Turbidity can also vary seasonally, as erosion rates, algal production, and sediment settling rates vary with factors such as temperature and snow cover. Therefore, collecting turbidity grab samples over a wide range of flows and seasons is better than having no site-specific turbidity information at all. Grab samples can be used to fit rough models of how turbidity varies with flow, season, and temperature.

Trout biologists have a tendency to focus on summer conditions, especially in studying the effects of water temperature. However, physical conditions in other seasons can have greater effects on trout growth and survival than summer conditions (Railsback and Rose 1999). It is important to assemble reliable data for the full year.

16.2. Selecting Reaches

Selecting the number, location, and size of study reaches is a critical step in applying InSTREAM. The following considerations provide some guidance, although the potential effects of reach location and size have not been investigated in detail. Here, "reach" refers to the length of stream represented in the model, and "site" refers to the longer length of stream that the model is intended to represent. One or more reaches are often used to represent a site.

For studies intended to assess the effects of river management, including instream flow studies, the primary consideration in selection of study reaches should be that reaches adequately represent the habitat diversity of the entire site. The number of reaches can be chosen to include any major differences in channel morphology within the site. The location and length of each reach should be chosen to include all the major habitat types (pools, riffles, cascades, etc.) and possibly also cover features (wood, undercut banks) in roughly the proportion they occur in the full site.

Spawning habitat should be considered in reach selection. If spawning habitat is relatively rare, the reach may need to be expanded or moved to include a representative amount of it. Because InSTREAM does not incorporate immigration, a site with little spawning habitat may exaggerate the variability of annual reproduction.

Habitat should not be excluded from the study reach because of complex hydraulics (section 16.5.2, section 16.6). Any errors in modeling complex hydraulics are likely to be less important than those that would result from ignoring important habitat.

Similarly, reaches (especially for instream flow assessment) should not exclude habitat types that are common but believed unimportant for trout. If, for example, a study site includes extensive exposed, shallow pools where trout are rarely observed, the pools should still be fully represented in the modeled reaches. Leaving out common habitat types, even if rarely used by trout, could bias study results in unpredictable ways.

Overly large reaches have computational disadvantages. The computer time required to execute model runs increases with both the number of fish and the number of cells, so it increases rapidly with the size of the study reach. Computation time can become an important consideration for reaches with thousands of fish and many hundreds of cells.

On the other hand, results from small reaches with low numbers of fish can be vulnerable to random effects. (Remember that InSTREAM does not represent emigration or immigration.) This limitation is especially important if one of the study objectives is to evaluate a population's persistence—its resistance to extinction. Although the relation between reach size and frequency of extinction in InSTREAM has not been explored formally, extinction is more likely for small reaches containing few fish. Stochastic effects on spawning may also cause high (and probably unrealistic) variability in simulation results. The number of adults that spawn is modeled stochastically, and one often-important cause of redd mortality (scouring) is also highly stochastic. This stochasticity tends to be higher at smaller reaches with fewer trout or little spawning habitat.

The reaches to which InSTREAM has been applied by the authors range in length from 75 m for a small, first-order creek to about 200 m for a third-order stream 5 to 10 m in width at base flow, to about 1000 m for a large mainstem river 20 to 40 m in width at base flow.

16.3. Transects and Cell Boundaries

Once a study reach is located, the next step is to lay out the boundaries that define its cells. (See section 16.6 for a note concerning alternatives to the PHABSIM-based approach discussed here.) The goal of this step is to produce a coarse map of the reach's habitat, capturing the full range of habitat types while trying to otherwise minimize the number of cells. The hydraulic models used to prepare input for InSTREAM are one-dimensional, assuming flow is always in one direction, variable among cells along cross-stream transects, and homogenous within cells. Consequently, the habitat map built from the field data can be envisioned as a series of parallel transects—rows of cells across the channel—going from downstream to upstream (fig. 47).

The importance of focusing on habitat instead of hydraulic precision when selecting transects considerations deserves reemphasis. The choices of how many transects and cells, and where to put the boundaries between them, should not be based on how these choices affect calibration of the hydraulic model. Representing habitat accurately and otherwise minimizing the number of cells is of far greater concern; the hydraulic modeling methods recommended in section 16.6 are designed to accommodate transects with small numbers of cells (even only two or three cells per transect that are submerged at low flow) in complex habitat.

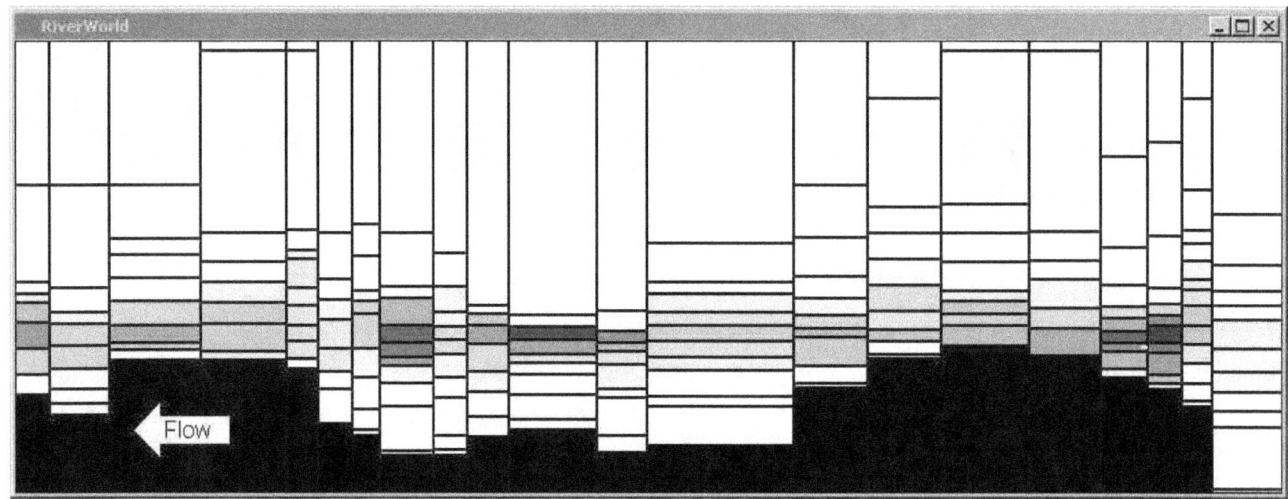

Figure 47—Transects and cells at the Little Jones Creek lower mainstem study site. Flow is from right (upstream) to left. Transect 1 is the most downstream row of cells (appearing on the figure as the leftmost vertical column of rectangles). Cell 1 on each transect is the leftmost (facing upstream; in the figure, cell 1 is at the top of each transect). The first and second cells on many transects are large, to represent the flood plain and to align the transect's thalweg (deepest cell) with those of neighboring transects. Cells are shaded by depth, with darker being deeper.

InSTREAM and PHABSIM use different definitions for transects and cells (see the habitat description conventions defined in section 3). In PHABSIM, a transect is a line across the channel through the middle of the habitat it represents; each cell on the transect extends both upstream and downstream from the transect. In InSTREAM, a transect is a row of rectangular cells across the river and transects are laid out by identifying the boundaries **between** sections of habitat, not by putting transects through the middle of a section of habitat. In other words, the field objective is to identify the coordinates of the black line segments in figure 47 that depict cell boundaries. It helps to remember that InSTREAM assumes the depth and velocity are uniform within each cell, so a transect is treated as a series of level rectangular cells (fig. 48).

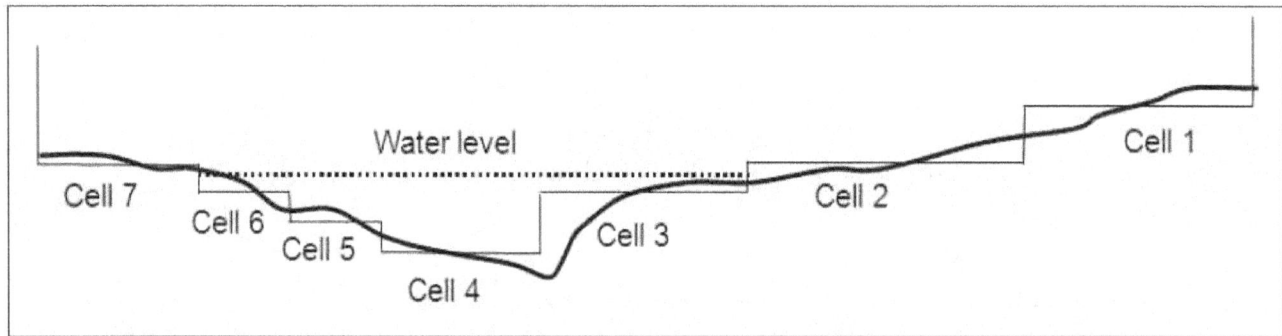

Figure 48—An example cross-stream transect, showing individual model cells along the transect and the actual channel cross section (heavy line).

Cell boundaries are defined using two numbers. First is the cell's upstream (X axis) coordinate: the distance between the downstream end of the whole reach and the upstream end of the cell. This value is the same for all cells on a transect. Second is the cell's across-channel (Y axis) extent: the distance from the right end of the transect (facing downstream) and the left side of the cell.

To lay out a reach's transects, start by identifying and marking the downstream end of the first transect, which is also the downstream end of the entire study reach. Then, moving upstream, identify and mark the upstream end of each transect. This boundary should be at a significant break in habitat conditions, such as the boundary between a pool and a riffle, or the downstream end of a cover feature such as a debris pile. It is also sometimes necessary to put a break between transects in the middle of a long, gradual change such as a gradually-deepening pool. The process is repeated upstream through the reach, marking boundaries between transects at the major changes in habitat longtudinally. The fact that rivers bend but InSTREAM uses a straight model of the habitat can require marking the transect boundaries on both sides of the channel and estimating cell length as the average distance between transect boundaries measured on the left and right side of the channel.

After the transect boundaries are established, the next step is to identify the boundaries between cells on each transect. It is very helpful to stretch two measuring tapes across the channel at transect boundaries. Then move across the channel between the two tapes and place cell boundaries at breaks in habitat such as rapid changes in depth or substrate type. It is also often necessary to break cells in the midst of a long, gradual change in depth.

The general goal in laying out transect and cell boundaries is to maximize the habitat variability between cells, while minimizing habitat variability **within** cells. Keep the following guidance in mind:

- Do not attempt to intentionally include or exclude any particular type of habitat. Instead, the goal is to accurately represent all the habitat types present.

- Remember the model's minimum spatial resolution of approximately 1 m^2 (roughly, the area used for feeding by an adult trout). Do not try to capture features occupying less area than this resolution. Do not, for example, attempt to place cell boundaries to capture small velocity breaks behind boulders. These are instead represented via the cell's velocity shelter variable (section 16.4). Avoid making cells less than 1 m in either length or width.

- Modeling considerations do not place upper limits on transect spacing or cell width. Large patches of uniform habitat should be represented using few cells.

If the study potentially will simulate flood flows, it is important for the transects to include cells representing any flood plains. In InSTREAM, as in real streams, flood plains can provide critical refuge from flood flows. If flood plains are extremely wide, it may not be critical to include their entire width, but only enough to provide refuge habitat.

Finally, once the cell dimensions are entered and running successfully in InSTREAM, it will be desirable to adjust their Y coordinates so that the deepest parts (thalwegs) of adjacent transects line up with each other. The reason for this adjustment is to minimize the error in distances between cells introduced by the one-dimensional depiction of stream geometry; exaggerating the distance between cells can greatly reduce the habitat selection options available to small fish (see the discussion of habitat selection in chapter 2). This thalweg adjustment can be done by adding artificial cells or adjusting the Y coordinate of their first cell, in such a way that the Y coordinate of the deepest cell is approximately the same in adjacent transects (fig. 47).

16.4. Cell Habitat Variables

Several habitat variables must be estimated for each cell. These variables (defined in section 5) represent the availability of velocity shelters for drift feeding, the distance to hiding cover, and the availability of spawning gravel in each cell. None of these variables can be defined very precisely, and variation in their values with flow and fish characteristics is ignored by InSTREAM. Therefore, the best way to evaluate them is simply by visual estimation by observers experienced with trout feeding, hiding, and spawning behavior. Recommendations by Railsback and Kadvany (2008) for judgement-based evaluation of stream habitat are applicable:

- Use more than one observer. Observers should be sufficiently experienced to have a good mental model of the kinds of habitat trout use for drift feeding, hiding, and spawning.
- Discuss and define in advance how the cell habitat variables are interpreted for the study site, especially what size fish are represented and what kind of cover they use for hiding and feeding.
- Discuss and agree on the values for each cell in the field, instead of recording separate values and averaging them later.
- Focus on estimating cell-average values, recognizing that there is no single "right" value.

Estimation of these variables may best be conducted during low flows when more of the stream channel can be waded and observed. However, at extremely low flows many of the cells can be dry, which introduces additional uncertainty into the estimates. Estimation of these values can be challenging in large and turbid streams.

16.5. Cell Hydraulics

Once cell boundaries are established, input to calibrate cell hydraulic simulations (section 16.6) must be collected. (See section 16.6 for a note concerning alternatives to the PHABSIM-based approach discussed here.) The necessary hydraulic data are characteristic bed elevations for each cell (measured once); and water surface elevation (WSE) for each transect and mean velocity for each cell, measured at several different flow rates covering a wide range. Measurements of WSEs during high and flood flows are important for studies that will require simulation of high flows. The term "calibration data set" refers to measurements of WSE and cell velocity, at one flow rate.

Although PHABSIM hydraulic models are recommended for hydraulic simulation, conventional PHABSIM data collection methods are not recommended. Users should understand the hydraulic simulation methods recommended in section 16.6 before collecting the cell hydraulic data. It is especially important to understand that, for InSTREAM, we are not interested in measuring or modeling streamflow. Instead, we are interested only in modeling the characteristic velocity and depth of each cell at any flow. Therefore, there is no need to follow stream gaging techniques such as measuring velocity at 20 points across each transect.

16.5.1. Selecting calibration flows—

A key decision in collecting cell hydraulic data is how many calibration data sets to collect, at which flows. A weakness of the hydraulic simulation methods is that they do not capture complexities such as the formation then disappearance of eddies and reverse velocities as flow increases, except to the extent those complexities are captured in the field data used for calibration. Therefore, it is important to collect several calibration data sets covering the range of flows to be modeled with InSTREAM.

It is important to model cell depths and velocities reasonably well at high flows, including the highest flows to be simulated. If a study site is downstream of a large reservoir so that flood flows are extremely rare, it may not be important to model such high flows. However, at most sites, such flows occur with sufficient frequency that they can neither be neglected nor modeled poorly. Without calibration data from at least one high flow, both depths and velocities are subject to major errors.

These errors may include predicting widespread extreme velocities and, consequently, erroneous prediction of severe fish mortality. At most sites, it is difficult and dangerous to measure cell velocities during high-flow events. However, high flows can be modeled with reasonable accuracy with only the WSEs at each transect as input (section 16.6).

Currently, most applications of InSTREAM have been in coastal California, where seasonal flow variation is extreme. Even at such sites, adequate hydraulic simulation has been attained with full data sets from three flows, plus one set of water surface elevations near or above bank-full flow. The three full-calibration data sets can be obtained at flows (1) approximately double the lowest flow to be simulated (assuming that the lowest simulated flow will be extremely low and therefore unlikely to be observed in the field), (2) near the highest flow at which measurements can be taken safely, and (3) a flow about halfway between the other two.

16.5.2. Collecting data—

The first kind of cell data needed for hydraulic simulation is a characteristic mean bed elevation for each cell. The average depth of the cell will be calculated by InSTREAM by subtracting this characteristic bed elevation from the modeled WSE. For cells with relatively uniform bed slopes, the characteristic bed elevation can be estimated as the elevation near the cell center. However, for cells with elevation peaks, troughs, many large boulders, etc., it may be necessary to survey elevation at several points and average the values. Bed elevations are most easily measured during low flows.

It is essential to install stable benchmarks so that bed elevations and the WSEs for all calibration data sets can be measured using the same baseline datum. The same datum must be used for all transects so that longitudinal slope throughout the study reach can be estimated (section 16.7.2).

The second kind of cell hydraulic data is the calibration data sets collected at each of several flows. The following procedures have proven useful for collecting cell data for hydraulic model calibration.

- Using surveying equipment, measure the WSE for each transect. If WSE varies from one side of the channel to the other, measure it on both sides and use the average. (The hydraulic models do not allow WSE to vary among cells on a transect.) WSEs must be measured from the same benchmark datum at each flow, so a relationship between flow and WSE can be developed.
- For each cell, select a location with velocity characteristic of the cell average, and measure velocity there. In fairly simple hydraulics, this location

can simply be the center of the cell. In complex hydraulics, it may be useful to select several locations and average the measurements from each location. Remember that measurements are intended to produce an average or typical velocity, not the exact velocity at any one point. The cell velocity should reflect conditions outside the influence of any velocity shelters in the cell. The effect of velocity shelters is modeled as a reduction from the cell-average velocity (see chapter 2). It is not important to take velocity measurements at the same location at each of the flows.

- It is a good practice to record the depth as velocity measurements are made, even though this information is normally not used for calibration. This practice allows detection of any major changes in bed elevation between site visits.

- Velocities are always considered positive, even if the water is flowing in the opposite direction of the channel average. (Technically, we are measuring water "speed," a scalar, not velocity, which is a vector that implies direction in addition to speed.) All velocities are considered positive because local flow direction is assumed to have no effect on fish feeding and mortality risks. (Remember, we are not trying to model the stream's total flow, only local velocities.) Some particularly complex cells could include water going in a variety of directions.

- The stream flow rate must be known for each calibration data set. If flow is not available from a reliable gage, then it must be measured. The cell hydraulic measurements described above cannot be relied upon to estimate flow accurately, so a separate measurement is required. The measurement should use standard flow measurement techniques, including selecting a special transect where velocities are relatively simple (which can be outside the reach being modeled) and using a large number of depth and velocity measurements across the transect. If flow is expected to increase or decrease during field data collection, then flow should be measured both before and after cell velocities are measured. Then, the flow occurring during measurements at each transect can be estimated and used for calibration.

- The third kind of data collection that can be used is measurement of WSEs only during high flows. This is done by simply surveying WSEs that represent the average for each transect. However, the flow during the WSE measurements must also be known. If there is no gage, it is necessary to estimate flow as well as possible, e.g., by measuring it from a bridge.

16.6. Hydraulic Simulation With RHABSIM

InSTREAM was designed so that hydraulic simulation—modeling the depth and velocity in each cell, at each daily flow rate—is separate from the fish model. This approach avoids the need to duplicate existing hydraulic models and allows use of the most suitable hydraulic models for each application. This section assumes users are familiar with the PHABSIM hydraulic models commonly used for instream flow studies, or will become familiar with them through the training materials and classes provided by the organizations that distribute the models. It is very important to understand the detailed assumptions used in these models, and especially to understand how their use as recommended here for InSTREAM differs from how they are typically used for PHABSIM.

Potential InSTREAM users should be aware of alternatives to PHABSIM/ RHABSIM for hydraulic simulation. Two alternative versions of the InSTREAM software have been developed. One simply uses lookup tables of WSE and velocity versus flow from a simple input file that could be generated from any source, including field measurements. The second—version 4.3, or InSTREAM-2D—uses a two-dimensional hydrodynamic model for hydraulic simulation, retaining the full two-dimensional geometry of the study reach. This version is especially suitable for large rivers, as some hydrodynamic models do not work well in shallow water. However, InSTREAM-2D cannot represent multiple linked reaches, only a single reach. See the InSTREAM Web site at http://www.humboldt.edu/~ecomodel for information.

The lookup table information used by InSTREAM (velocity and WSE for each cell, at each of many flow rates over a wide range; section 5) could potentially be generated by many different kinds of hydraulic models. However, the InSTREAM software is currently designed to directly import these tables from the RHABSIM© river simulation package, a commercial version and enhancement of the PHABSIM hydraulic models available from Thomas R. Payne & Associates (Arcata, California; 707-822-8478; http://www.northcoast.com/~trpa/). RHABSIM is currently available at no cost and is widely used. The package includes the options described below that provide at least a rudimentary ability to simulate complex velocities at a spatial resolution appropriate for trout, and provides an output file that acts as lookup table input to InSTREAM. (PHABSIM for Windows, another version of the PHABSIM models, is in the public domain and free but does not provide an output file that can be used directly by InSTREAM; see http://www.fort.usgs.gov/products/ software/PHABSIM/PHABSIM.asp. Were there demand, PHABSIM for Windows and InSTREAM could readily be modified to work together.)

The RHABSIM methods recommended here are designed to accommodate complex hydraulic habitat as well as possible. We recommend users ignore some conventional PHABSIM calibration procedures, especially paying attention to how well the simulated depths and velocities in each cell along a transect add up to match the given flow rate. Users should ignore "velocity adjustment factors" and should not let the hydraulic model adjust cell velocities to fit the given flow rate. Remember that the objective is not to model the total river flow across a transect but to model the depth and velocity in each cell as well as one can, without "calibrating" away the natural complexities important to fish.

One important natural hydraulic complexity is the tendency of a cell's velocity to change nonlinearly with flow and even to decrease as flow increases (e.g., as eddies form and then disappear with increasing flow). The regression ("IFG4") method of modeling velocities should not be used because it assumes a log-linear relationship between velocity and flow. The best approach is to use the "one velocity calibration" approach: modeling how velocity varies from the measured velocity at flows around each calibration flow.

Using the "one velocity calibration" approach, users select one of the calibration data sets, which is then used by RHABSIM to set the parameters to calculate cell velocity from depth. Then the user selects a set of flow rates for which cell hydraulics are calculated using those parameter values. This process is repeated with each set of calibration data used to simulate specific flows in different ranges. InSTREAM uses results from all these flow rates to build its lookup table of cell velocity and WSE versus flow.

Users must therefore decide in advance what range of flows to simulate using each calibration flow. Considerations include:

- The total range of flows simulated using all calibration data sets needs to include the lowest and highest flows expected to be modeled in InSTREAM. Although InSTREAM is designed to extrapolate below or above the range of flows included in its cell velocity and WSE lookup tables, results will be more accurate if the full range of flows is simulated in the hydraulic model.

- Many flows should be simulated instead of having large gaps between simulated flows. InSTREAM estimates cell velocity and depth by linearly interpolating among the simulated flows. The accuracy of this interpolation will be much better if many flows are simulated. The relationships between flow and depth or velocity are typically fairly linear over small changes in flow. Typical hydraulic simulations for InSTREAM include four calibration flows, with 10 to 20 flows simulated for each calibration flow.

- Hydraulic simulations are generally assumed to be better if simulated flows are no more than 1.5 to 2 times their calibration flow.

A hypothetical application of InSTREAM to the Mad River near Arcata, California, provides an example. Historical flows from 1975 to 1995 are to be used as input. The range of flows occurring in this period is 0.003 to 1040 m^3/s, with 25 percent of daily flows being less than 1.3 m^3/s and 50 percent of flows less than 6 m^3/s. Assume that full calibration data sets were collected at 1.0, 8.0, and 25 m^3/s, and a set of high-flow WSEs observed at 200 m^3/s. Table 26 illustrates the simulation flows that could be used for this example.

Table 26—Example simulation flows

Calibration flow	Simulated flows
	Cubic meters per second
1.0	0.003, 0.006, 0.01, 0.02, 0.04, 0.06, 0.08, 0.1, 0.2, 0.3, 0.4, 0.5, 0.6, 0.7, 0.8, 0.9, 1.0, 1.2, 1.4, 1.6, 1.8, 2.0
8.0	2.5, 3.0, 3.5, 4.0, 4.5, 5.0, 5.5, 6.5, 7.0, 7.5, 8.0, 9.0, 10, 11, 12
25	14, 16, 18, 20, 24, 26, 28, 30, 35, 40, 45, 50
200	60, 70, 80, 90, 100, 120, 140, 160, 180, 200, 250, 300, 400, 500, 600, 700, 800, 900, 1000, 1050

Specific steps in using RHABSIM for full calibration data sets are:

1. Import the calibration data into RHABSIM. The "Text X/Y/Vel" facility is useful. All calibration data sets, including WSEs measured at high flows, should be included in one file.
2. If the field data were input in English units, use the RHABSIM facility to convert the file to metric.
3. In the RHABSIM "HYDSIM" menu, select "WSLs" and "Select WSL Methods." Set the WSL method to log-log regression.
4. In the WSLs, "Calibrate WSLs" menu, check the log-log regression relation between WSE and flow for each transect.
5. Also in the WSLs, Calibrate WSLs menu, select the "Dual SDR" method. This method tells RHABSIM not to adjust cell velocities to reproduce the flow rate across the transect. (It may be necessary to first go to the "Edit, Stage/Discharge/SZF" menu item and execute the "Calc Flow" option.)

6. In the "VELS," "Parameters" menu, set the velocity calibration method to "1-vel calibration." This tells RHABSIM to calibrate velocities for a range around only one of the calibration data sets. Set the "VELSET" to one of the calibration data sets.

7. Go to the "Edit" menu (still under HYDSIM), and select "Calibration Flows." Enter all the flows to be simulated using the calibration data set's flow as a basis. These must be in increasing order (this is a requirement of InSTREAM, not RHABSIM).

8. Under the VELS menu, use the "Roughness worksheet" and its graphics to review the simulated velocities at each transect. Cell velocities can be calibrated in this menu.

9. When velocities are acceptable, use the VELS menu's "View Vels" option to write the output file that is used by InSTREAM. An example of this file is below.

Steps 6 through 9 of this process are repeated for each calibration data set. InSTREAM imports one of the "View Vels" files for each of up to five ranges of calibration flows.

Hydraulic simulation is slightly different for high calibration flows at which only WSEs were measured. In step 5, do not select the "Dual SDR" method, and in step 6 set the velocity calibration method to "Depth calibration" instead of "1-vel calibration." These options cause a cell's velocity to be calculated from its depth; then the velocities across a transect are adjusted to match the total flow. (In other words, the total flow is allocated among cells by assuming deeper cells have higher velocity.) Even though this velocity adjustment is not desirable for full calibration data sets, it is desirable for high flows calibrated only with WSEs. The method used to estimate cell velocity from depth is extremely simplistic and inaccurate without velocity adjustment, and at high flows velocity fields tend to be less complex.

Following is an example hydraulic data file produced by RHABSIM. This example shows only one transect, with simulated flows ranging from 0.003 to 0.012 m^3/s. "STATION" refers to distance from the right bank (facing downstream), each station referring to one cell in InSTREAM. "ELEV" refers to the cell's bed elevation. Depth at a cell is calculated by subtracting ELEV from WSL.

```
Little Jones Creek, Weejak Creek trib ASCII input file
SFR 3/28/00

Velocity ALGORITHM: Roughness (Manning's N)
Use Given N's? Yes

VELOCITY table for XS # 1 T
WSLs based on Log/Log Regression
VELs based on 1-vel calibration
Calibration Set Used: 2
VEL Method: Centered over vertical
```

	FLOW:	0.003	0.004	0.006	0.008	0.01	0.012
	WSL:	93.04	93.05	93.07	93.08	93.09	93.09
Wet Cells:		3	4	4	4	4	4

STATION	ELEV						
0.0	97.00						
6.7	93.66						
8.9	93.04		0.00	0.00	0.01	0.01	0.01
9.9	92.88	0.01	0.01	0.02	0.02	0.02	0.03
10.9	92.81	0.11	0.13	0.15	0.17	0.19	0.21
11.9	92.74	0.01	0.02	0.02	0.02	0.02	0.02
12.8	94.55						
18.0	97.00						
Average Vel:		0.05	0.04	0.05	0.06	0.06	0.07

16.7. Reach Parameters

This subsection discusses estimation of values for several habitat parameters that depend on hydraulic conditions at the reach scale.

16.7.1. Maximum flow for spawning—

InSTREAM assumes trout will not spawn on days when the flow is greater than the parameter *habMaxSpawnFlow*. The only good basis for evaluating this parameter is judgment based on observation of conditions during relatively high flows. The parameter is included in InSTREAM to implement the assumption that fish will not spawn when flows are so high that redds would likely be scoured. Therefore, *habMaxSpawnFlow* should be a flow below that at which widespread movement of spawning gravel appears imminent. If spawning gravel is primarily in gravel beds or bars, then the scouring model (below) may be useful for estimating *habMax-SpawnFlow*. But the scouring model will not be useful if spawning gravel is scarce or occurs only in pockets behind obstructions such as boulders.

16.7.2. Shear stress parameters—

Two parameters, *habShearParamA* and *habShearParamB*, are used to relate reach-average Shields shear stress to flow in the formulation for scouring and deposition redd mortality. The first decision users must make about these parameters is whether the scouring and deposition formulation is appropriate for their study reach. The formulation is designed for alluvial streams where spawning gravel beds or bars are fairly extensive. In reaches where spawning is primarily in pocket gravels behind obstructions, one might assume that scour processes can be approximated with the model's formulation because in such reaches, scour of redds probably remains a stochastic event that increases with streamflow and decreases with redd depth. Alternatively, scouring mortality can be turned off by setting the trout parameter *mortReddScourDepth* to a high value such as 10 000 cm.

The parameters *habShearParamA* and *habShearParamB* can be evaluated by using an equation commonly used to estimate Shields stress in rivers:

$$shearStress \ = \ \frac{\rho^{RS}}{\left(\rho_s - \rho\right)D}$$

where ρ is the density of water (1.0 g/cm^3); R is the reach-average hydraulic radius (cm); S is the reach-scale energy slope (dimensionless); ρ_s is the density of sediment, approximated as 2.7 g/cm^3; and D is the median substrate particle diameter (cm). The hydraulic radius R can be approximated as the average depth and S can be approximated as the average water surface slope, using data collected for hydraulic model calibration at several different flows (section 16.5). Then logarithmic regression of *shearStress* versus flow produces the values of *habShearParamA* and *habShearParamB*.

Evaluating Shields stress with hydraulic model calibration data, however, may introduce significant errors if the study reach has numerous sharp bends and obstacles. Including these in the reach over which Shields stress parameters are evaluated would overestimate shear stress and the potential for scouring. This potential problem can be circumvented by estimating the relation between Shields stress and flow at a reach that is relatively straight and obstacle-free but has a slope and substrate diameter similar to the study reach. This Shields stress measurement site could be a part of the InSTREAM study reach or a separate site nearby on the same stream.

16.8. Barriers

Barriers to trout movement can be represented in InSTREAM (section 5). The only field datum needed to characterize a barrier is its location, designated as the distance upstream from the downstream end of a reach.

16.9. Trout Census Data

It is possible to use InSTREAM without any site-specific data on the trout being modeled, but census data should be considered essential for typical applications. Uses of census data include specifying the initial population, developing the trout length-weight parameters, and calibrating the model.

Initializing the model requires the user to provide the abundance and length distribution of each age class when the model starts (section 8). This requirement means that census data will be more useful if broken out by the age classes used by InSTREAM, and are collected at the same time of year that model runs are to start. However, after simulations run for several years, the assumed initial population abundance and age distribution has little effect (see section 15.3).

Field data on length and weight of individual trout can also be used to evaluate the two parameters *fishWeightParamA* and *fishWeightParamB* (section 6). These parameters represent the length-weight relation for trout in good condition, which can vary among sites. If data on sufficient individuals are available at the study site, they can be analyzed to identify individuals in good condition. A standard condition factor (e.g., $100{,}000 \times$ weight $\div$ length cubed) can be calculated for each measured fish. Then the fish in good condition (e.g., with condition factor in the top third of all values) can be identified and used (via log-log regression of weight versus length) to find values of *fishWeightParamA* and *fishWeightParamB*. (Values of these parameters can also be obtained from the literature, e.g., Carlander 1969.)

17. Calibration

Calibration of a model refers to adjusting parameter values to improve how well model results reproduce observed data. For simpler models, calibration is often considered the most important step in modeling. The closeness of the calibrated fit of model results to data is often considered the most (or only) important measure of a model's accuracy. There is an extensive literature on calibrating simple models to data (e.g., Hilborn and Mangel 1997).

For IBMs, however, traditional calibration to data is less important than developing confidence that the model's mechanisms and processes are reasonably accurate. A complex IBM can be forced to reproduce an observed data set even if its internal mechanisms are very poor, so calibration is not meaningful unless preceded by analysis and testing of the model's mechanisms and processes (as done for InSTREAM in chapter 2). On the other hand, experience with InSTREAM confirms that well-tested mechanistic models can be used for many kinds of analyses without the need for site-specific calibration. Models with detailed representation of mechanisms can be much less dependent on calibration.

Many consumers of model results (e.g., decisionmakers) automatically consider a model's ability to reproduce observed data as the primary measure of the model's usefulness. Therefore, many users of InSTREAM will need to conduct and document some kind of calibration. On the other hand, application of the model to management decisions must often be made without extensive calibration. Because InSTREAM is a long-term population model, calibration requires an extensive time series of population observations. These data—or the time and resources to collect calibration data—will not be available for many applications.

In general, users of InSTREAM are encouraged to compare model results to data representing conditions similar to those modeled. Parameters can be adjusted to reproduce observations if there are indicators of which processes need adjustment. However, users should keep in mind that they could, in forcing the model to reproduce a unique (or uncertain) observation, change a parameter value from one that is reasonable in general to one that represents a relatively rare specific condition. Two particular problems (addressed further below) should also be kept in mind:

- One output can be calibrated by adjusting several different parameters. For example, the abundance of age 0 trout could be calibrated by adjusting several parameters that control how many adults spawn or by adjusting the level of predation by other fish.

- Results of InSTREAM are at least as sensitive to the input data that depict site conditions as they are to parameter values. Therefore, data observed under one set of conditions (e.g., at a nearby but not identical reach, or before versus after a flood that re-shaped the channel) may not be useful for calibrating the model to another set of conditions. Also, error in input data could have strong effects on calibration.

Users are reminded to consider the robustness issues discussed in section 14.4 while conducting calibration: simpler calibration methods may be appropriate if a few simulation experiments show that study conclusions based on InSTREAM are robust to variation in key calibration parameters.

Two kinds of calibration are discussed below: relatively simple manual adjustment of parameters and systematic calibration using statistical and graphical analysis. Both of these approaches require selecting, in advance, a limited number of parameters to adjust.

17.1. Calibration Parameters and Outputs

It is impossible to calibrate all of InSTREAM's many parameters to all of the different kinds of output it produces. Hence, the first calibration step is to select which parameters to calibrate, and which outputs to use as indicators of calibration quality.

The choice of model outputs to use in calibration is usually highly constrained by available data. The model can only be calibrated using outputs that directly correspond to values measured in the field, and meaningful calibration requires several years of data. Typically, the first year's field observations are used to initialize InSTREAM, and calibration based only on the next year or two of a simulation is questionable because results can be affected by initial conditions (section 15.3). Therefore, data from several additional years are needed for a convincing calibration. Data from several observations within a year are likely to be more useful than one observation per year, but data from multiple years is desirable to ensure that long-term dynamics are reasonable. (Remember that InSTREAM ignores some seasonal processes, especially variation in food production, so calibration at an annual time scale is more meaningful than calibration over shorter times.)

A typical stream trout census estimates the density of trout in several age classes, and the size distribution within each age class. With these field data, InSTREAM can be calibrated by fitting model predictions of abundance and mean length (or weight) of each age class, and possibly also the variation in size. Other census techniques may provide fewer or different observations. For example, snorkel surveys typically produce less information (or less certain information) on size and age. Spawning surveys may be used (cautiously, as these data are often quite uncertain) to calibrate the number of redds produced and when spawning occurs.

In choosing which InSTREAM parameters to adjust in calibration, we can use one lesson from traditional, simple ecological models: calibrating fewer parameters increases our confidence in their calibrated values. Trying to adjust too many parameters reduces our ability to find the best range of values for each, so it is generally recommended to adjust as few parameters as possible to fit the available observations. The parameters most suitable for calibration are those to which model results are highly sensitive (section 15.2) and for which there is little basis, other than calibration, for selecting values. These two criteria, and experience with InSTREAM, indicate four parameters that are especially suited for calibration:

- Concentration of drift food (*habDriftConc*): This parameter strongly affects growth of all age classes. Data on actual drift availability are rarely available and, when available, notoriously variable and uncertain. (Calibrated values of *habDriftConc* may be quite different from observed drift data because this parameter also captures effects of several processes that are neglected in InSTREAM, e.g., variation in prey size and energy content, and spatial and seasonal variation in drift production.)

- Survival of terrestrial predation (*mortFishTerrPredMin*): Terrestrial predation is typically a dominant source of mortality for all but the smallest trout, and actual predation rates are almost never known.

- Survival of aquatic predation (*mortFishAqPredMin*): Small juvenile trout are highly vulnerable to predation by other fish, but the actual risk is highly variable and difficult to quantify. Therefore, this parameter is particularly appropriate for calibrating the abundance of juveniles. (However, the sensitivity analysis discussed in section 15.2 did not find *mortFishAqPredMin* to have strong effects on predictions of adult trout production.)

- Production of stationary food (*habSearchProd*): Juveniles are the only trout that consistently use search feeding in InSTREAM, although juveniles often also use drift feeding. This search food parameter is therefore useful for calibrating differences in growth between juveniles and larger trout.

- Regeneration distance for drift food (*habDriftRegenDist*): This parameter affects the total availability of drift food per cell, which can limit how many trout can occupy each high-quality cell. Consequently, it has strong effects on populations. This parameter could be calibrated by attempting to match observed densities of trout in high-quality habitat.

InSTREAM will undoubtedly be used at sites where few or no calibration data are available. Applying InSTREAM in the absence of calibration data is not necessarily discouraged. If stream management decisions must be made at sites where data are lacking, using a more mechanistic model such as InSTREAM may be more appropriate than using simpler, more data-dependent, models—especially when the simpler models would be calibrated with data such as habitat "preferences" from other sites. In the absence of the kinds of data discussed above, model users can at least adjust key calibration parameters to ensure that basic outputs such as adult size and density are reasonable. Finally, even 1 year of data can be useful in calibration. Particularly if observations are not preceeded by unusual conditions, model output should produce results in the range of observed data.

17.2. Manual Calibration

"Manual" calibration refers to fitting InSTREAM's results to observed data by adjusting one parameter at a time. The variable can be adjusted simply by trying several different values and trying to zero in on a good value, but it is typically more efficient and rigorous to use a sensitivity analysis approach (section 14.2.2), running the model several times using a wide range of values. The sensitivity analysis approach illustrates how the model responds to the calibration parameter and how much of the variation in results is due to stochasticity. Calibration model runs must be long enough so that results are not affected strongly by initial conditions (section 15.3).

For several data sets, at least a coarse calibration of InSTREAM has been achieved using the following manual steps.

- Fit size (length or weight) of age 1 and older trout by adjusting *habDriftConc*.
- Fit abundance of age 1 and older trout by adjusting *mortFishTerrPredMin*.
- Fit abundance of age 0 trout (if data are available) by adjusting *mortFishAqPredMin*.
- Fit size of age 0 trout by adjusting *habSearchProd*.

After each of these steps is completed, of course, it is necessary to check whether the previous steps have been affected. Small effects should be expected; for example, adjusting *habSearchProd* is likely to have some effect on adult size, but perhaps not enough to require recalibration of *habDriftConc*.

17.3. Systematic Calibration

"Systematic" calibration refers here to using systematic simulation experiments and analyses to find parameter values meeting specific, quantitative calibration criteria. This approach is appropriate when extensive field data are available and it is especially important to establish the model's ability to reproduce the field observations.

This discussion of systematic calibration is based on the work of Cunningham (2007), who developed systematic calibration methods and applied them to the lower mainstem reach of Little Jones Creek. The data set included annual fish censuses from 6 years, the first of which was used to initialize the model. In addition to choosing which model outputs to compare to which data, and which parameters to adjust, this approach requires the following steps.

First is choosing a statistical measure of model goodness of fit to the data. Cunningham used the sum of squared deviations (SSD) between model results and observed data. For example, to calibrate age 2+ abundance, the model's output for age 2+ abundance was subtracted from the observed abundance, and this difference squared. One such squared deviation was calculated for the one day of each year when field observations were taken. The sum of the squared deviations from each year was the overall indicator of model fit; lower values indicate better calibration.

The second step is to identify a range of plausible values for each calibration parameter. It is usually best to use a wide range of parameter values, broken into relatively few values. For example, after preliminary experiments indicated that *habSearchProd* had little effect, Cunningham used the parameter values in table 27.

The third step is to execute the model for all combinations of parameter values. (For the parameter values in table 27, a total of 100 model runs was required to simulate all combinations.) The statistical measure of goodness of fit to data is then calculated for each model run. If desired, replicates of each of these scenarios can be generated to reduce the stochasticity in results.

Table 27—Parameters and values used in the systematic calibration by Cunningham (2007)

Parameter	Values
habDriftConc	2, 4, 6, 8, and 10×10^{-10}
mortFishTerrPredMin	0.980, 0.984, 0.988, 0.992, 0.996
mortFishAqPredMin	0.90, 0.93, 0.96, 0.99

The final step is to examine simulation results to find parameter combinations that produce good fit to data. One way to do so is via graphical analysis: plotting contours of the goodness of fit measure versus two of the parameters at a time. Graphical analysis is especially useful because the best model fit will be different for different outputs, and graphs are useful for finding regions of relatively good parameter values. Then parameter values that provide relatively good fit for as many outputs as possible can be identified.

Cunningham (2007) varied *habDriftConc* and *mortFishTerrPredMin* while keeping *mortFishAqPredMin* constant at 0.90. Figure 49 shows results for abundance and mean length of the age 0 and 2+ age classes; similar analyses were also conducted for age 1 trout. The contour plots show that—for three of the four outputs—the calibration is good (values of SSD are low) in a region where *habDriftConc* is 6×10^{-10} to 7×10^{-10} and *mortFishTerrPredMin* is 0.986

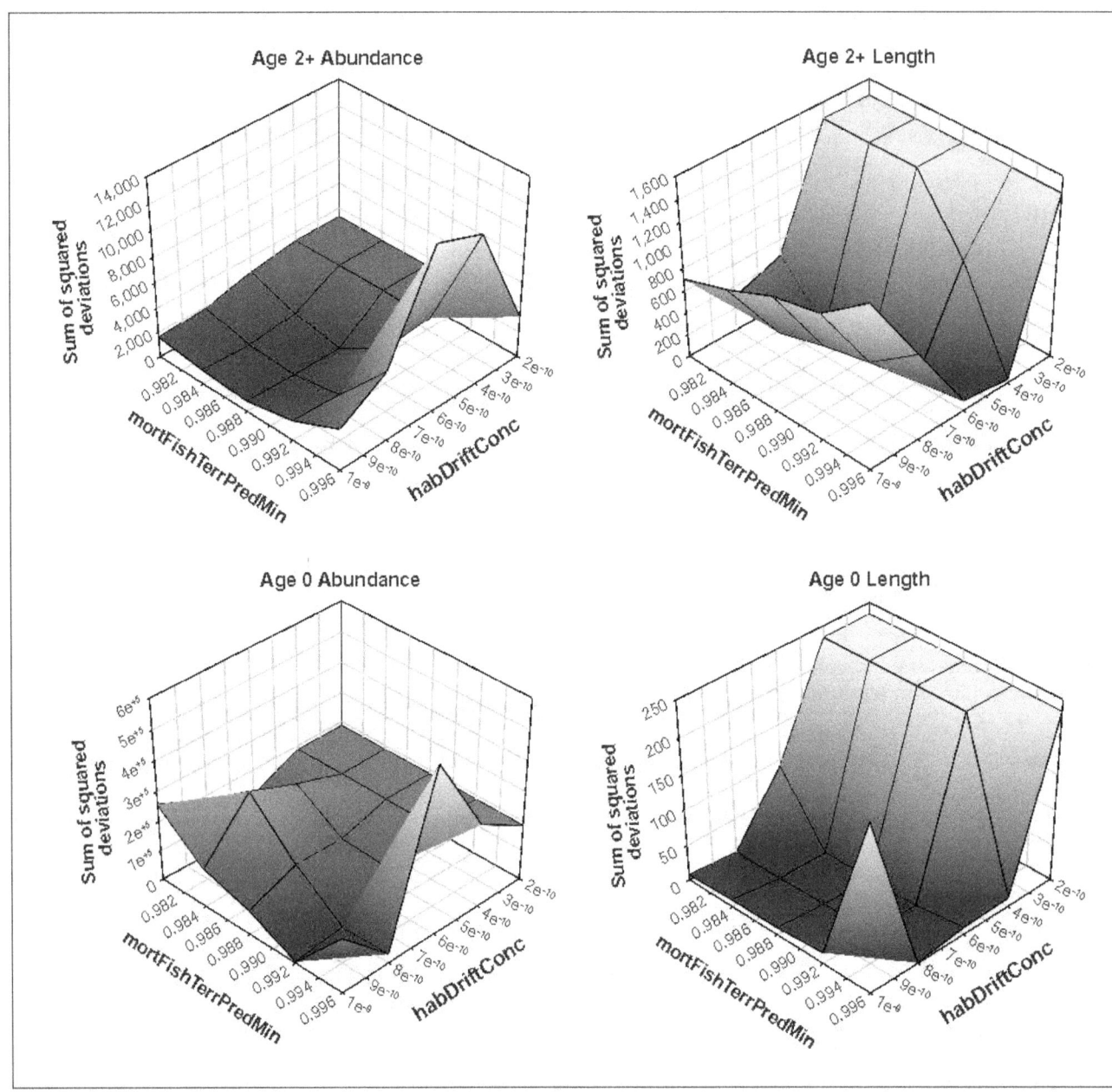

Figure 49—Plots of model goodness-of-fit (sum of squared deviations) vs. calibration parameters representing drift food concentration (X axis) and terrestrial predation risk (Y axis). Information from Cunningham (2007).

to 0.988. Calibration for age 1 abundance and length was also good in this region, and the calibration of drift concentration and terrestrial predation was robust to variation in *mortFishAqPredMin*. The lowest SSD values for age 0 abundance, however, occur at higher values of both parameters. Given the lower management importance of age 0 trout (and the higher uncertainty in census data for them), Cunningham chose calibration values of 6.5×10^{-10} for *habDriftConc* and 0.986 for *mortFishTerrPredMin*.

Chapter 4: Software Guide

18. Software Guide: Introduction and Objectives

18.1. Document Purpose

The software for individual-based models (IBMs) such as InSTREAM must not only implement the model's assumptions and equations but also provide a laboratory for observing, and conducting experiments on, the complex virtual ecosystem it creates. The purpose of chapter 4 is to provide the detailed information needed to execute InSTREAM on a computer and obtain the necessary outputs. Using InSTREAM requires software tasks such as installing the software, building and testing several kinds of input files, running the model, processing and analyzing output, and setting up automated multirun experiments. Many users will also choose to make simple revisions to the software to change which species are modeled or produce different kinds of output, such as "movies" of model runs or more detailed results.

18.2. Getting Started

We recommend that new users facing the task of installing InSTREAM and learning to use it:

- Read this section and pay special attention to section 18.8.
- Proceed through the instructions in section 19 while at the computer, conducting the installation and execution steps.
- See the troubleshooting guide in section 26 if problems arise.
- Use sections 20 through 25 as needed to set up and execute experiments.

18.3. Software License and Disclaimer

Like the Swarm simulation system it uses, InSTREAM's software is free and distributed under the GNU General Public License as published by the Free Software Foundation. In essence, this means that users are free to use and modify the software, but they must make the source code available to anyone they distribute the software to and cannot patent or make the software proprietary. The full GNU license is distributed with the software, in a file called "LICENSE." The InSTREAM software is copyrighted.

The InSTREAM software is distributed in the hope that it will be useful, but WITHOUT ANY WARRANTY; without even the implied warranty of merchantability or fitness for a particular purpose. See the GNU General Public License for more details.

18.4. Software Design Goals

Many IBMs of fish populations and river management problems have been built, but almost all were programmed from scratch and included many site-specific considerations in the software. Consequently, these models were useful only for one site and therefore unlikely to be widely used, tested, or maintained. Generally, IBMs have been criticized as being too complex to test adequately, lacking credibility as a regulatory decisionmaking tool, and costing too much (e.g., Bart 1995). These criticisms have arisen in part because too many IBMs lack software that (a) has adequate quality assurance, (b) facilitates testing and demonstration, and (c) allows models to be built, tested, and modified rapidly and inexpensively. The problems that result from modelers developing software from scratch are succinctly stated by Minar et al. (1996): too much of the modeler's time and money is spent on software instead of on model building and testing, and the software often lacks key qualities such as testability.

The software for InSTREAM was designed specifically to avoid these problems. The software is intended not just to implement a particular trout model, but to provide general, reliable, reusable, and adaptable software for fish IBMs. Goals in designing the software include:

- Allowing changes in formulation to be implemented easily. A well-organized, object-oriented software design helps address this goal, as does the choice of platform (below).
- Maintaining high-quality software and documentation. Undiscovered programming mistakes and undetected flaws in a model's formulation can be extremely expensive. Documenting software quality is very important for establishing the credibility of models for regulatory and scientific uses. The InSTREAM software provides a variety of tools to aid in testing. The software has been extensively tested for errors (below).
- Providing thorough and up-to-date documentation. Documentation measures in addition to this report include writing and commenting the source code to make it self-explanatory, and using automated version control to track changes in the code.
- Providing the graphical and file outputs necessary to allow users to observe and understand simulations, including the spatial distributions of habitat and the movement and growth of individual fish. Without such output, it is impossible to verify that a model is working as desired.
- Facilitating the kinds of model analysis and calibration discussed in chapter 3. The Experiment Manager is provided to support such analyses.

18.5. Software Platform: Swarm and EcoSwarm

The InSTREAM software is written using the Swarm simulation system, a library of software that essentially provides a programming language for IBMs (Minar et al. 1996). Swarm was originally developed by Chris Langton and colleagues at the Santa Fe Institute and is now distributed by the nonprofit Swarm Development Group. Swarm is probably the most widely used of several similar libraries for IBMs and agent-based modeling. Further information on Swarm is available from its Web site: http://www.swarm.org.

Swarm provides a software "framework" for IBMs: it includes a conceptual framework for designing models and their software, and software to provide functions common to all models: maintaining lists of objects, scheduling events, drawing pseudorandom numbers, observing results graphically, etc. Swarm is written in Objective-C, an object-oriented extension of the C programming language (NeXT 1995). The InSTREAM software is also written in Objective-C and provides the model-specific details not provided by Swarm—the characteristics and behaviors of the habitat cells, fish, redds, etc.

Swarm was selected as the software platform after an extensive review of alternative platforms for IBMs (see also Lorek and Sonnenschein 1999). The advantages of Swarm include:

- It is specifically designed for agent-based modeling, so it provides much of the model and graphics code needed for a powerful, flexible individual-based modeling system. This avoids the costs and potential errors that result from writing new code from scratch.
- Swarm has a talented, active, and diverse development team and user community. The users of Swarm include many scientists on the forefront of complex systems research (economists, sociologists, microbiologists, physicists as well as ecologists) and interaction with them is very beneficial.
- The software is portable among computers and operating systems, relatively easy to install and use, and free.

EcoSwarm is a library of Swarm-based software specific to ecological IBMs, maintained at Humboldt State University's Department of Mathematics. As InSTREAM and other models were developed, code for general functions was put in Objective-C classes designed and documented for easy re-use. This EcoSwarm software provides date and time management, reads and manages time-series input data and parameters, creates "breakout" statistical summary output files, and models survival probabilities. EcoSwarm is documented extensively at http://www.humboldt.edu/~ecomodel. All the EcoSwarm code needed for InSTREAM is packaged with it; EcoSwarm requires no additional installation.

18.6. Software Quality and Quality Documentation

The developers of InSTREAM make no warranty that its software is mistake-free. They have, however, conducted several kinds of tests to seek out and eliminate errors. Software quality measures (Grimm and Railsback 2005: chapter 8; Ropella et al. 2002) include:

- Using widely used library software (Swarm and EcoSwarm) as much as possible. Widely used software is less likely to contain undetected errors than is new code.
- Designing the software to be self-documenting: the primary goal in programming was not efficient computer execution but making the code easy for people to read and understand.
- Defensive programming to prevent run-time errors. The software includes numerous checks for run-time errors such as uninitialized variables, division by zero, and floating point overflow or underflow.
- Thorough code reviews. All source code has been reviewed repeatedly by at least one person other than the programmer who wrote it.
- Spot checks and visual checking. Graphical interfaces are used to check for errors producing obvious or widespread effects. Graphical interfaces often illuminate important problems that would otherwise go undetected.
- Comprehensive tests of key submodels against a separate implementation. Each major component of the software is programmed a second time, in a spreadsheet. Then output from the InSTREAM software, for thousands of cases, is imported into the spreadsheet and checked for differences. These tests have been conducted for cell hydraulics, reach habitat variables, food availability, feeding and growth, survival probabilities for fish and redds, and selection of spawning sites.
- Documentation of tests and revisions. Records showing that tests have been completed and how, and that the resulting revisions were implemented, are available upon request.

18.7. Contributors and Sponsors

The InSTREAM software was developed and is maintained under several projects. Initial development occurred in 1998 under the "Instream Flow Modeling" research project, Department of Mathematics, Humboldt State University, with funding from Pacific Gas and Electric Company and Southern California Edison Company. Followup funding for software documentation and testing was provided by EPRI, Electric Power Research Institute Inc., under agreement WO6953-02 with Lang,

Railsback & Associates (LRA). An earlier software guide was published by EPRI (Railsback 1999b). Additional development has been funded by EPRI under agreement EP-P3215/C1529 with LRA; and by the U.S. Forest Service, Pacific Southwest Research Station, under research joint venture agreement PSW-99-007-RJVA with LRA. Starting in 2003, software development and maintenance has been funded primarily by grant RD-83088601-0 from the U.S. Environmental Protection Agency's Science to Achieve Results (STAR) program.

A prototype of the software was designed and written in 1998 by Glen Ropella and Chris Langton of The Swarm Corporation. Since then, software development and maintenance has been conducted by Steve Jackson and Steve Railsback.

18.8. Software Guide Overview and Conventions

The software guide is designed so information appears roughly in the order it is needed in a model application. Instructions for installing and running the software are in section 19. Section 20 describes the input files and how they are assembled. How to add or change the trout species represented in InSTREAM is explained in section 21. Sections 22 through 24 describe output, including graphical interfaces, the standard file outputs that are almost always used, and more specialized outputs. Finally, section 25 describes the Experiment Manager, a tool for automating simulation experiments that use multiple model runs.

The software guide for earlier versions of InSTREAM (Railsback 1999b) included detailed description of the software itself: its classes, methods, data structures, etc. This detail is not included in this document because it is a distraction to most users. Programmers interested in making modifications beyond those described in this document should contact the InSTREAM developers for more detailed information.

Changes in the software are inevitable as its use continues. Users are encouraged to check the project Web site (http://www.humboldt.edu/~ecomodel) for updates or new versions of the software and this guide.

This document is generally directed to users of Windows operating systems. However, InSTREAM and the Swarm simulation system it uses are actually Unix-based software and can readily be used in Linux or other Unix-like operating systems. In Windows, InSTREAM runs in a Unix-like terminal window called Cygwin. Unix users can simply work within a Unix terminal window.

In the following instructions, commands that users are to type into the terminal window are distinguished by setting them in Courier font (`like this example`). Likewise, example file contents are in Courier.

19. Installing, Compiling, and Running InSTREAM

This section provides the basic instructions for installing the InSTREAM software, starting with installing the Swarm platform that InSTREAM runs under. The final subsection discusses several known limitations and problems.

19.1. Computer System Requirements

Swarm and InSTREAM can be operated in Microsoft Windows operating systems and in Unix-based operating systems such as Linux and Solaris. There are no particular hardware requirements, although most users will want a fast processor and ample random access memory. Current versions of Swarm (2.1.150 and later) execute simulations as fast in Windows as in Linux; previous versions (especially 2.1.1) executed considerably slower in Windows.

Current versions of InSTREAM (released after the start of 2005) are designed to work with Swarm version 2.2. Especially under Windows, InSTREAM cannot be assumed to work in Swarm 2.1.1.

For the rest of this section, instructions are specific to Windows. However, except for installing Swarm there is little difference among operating systems. Users of Unix-based systems should see the Swarm Web site for information on installing Swarm.

In Windows, Swarm works as a library within Cygwin. Cygwin (http://www.cygwin.com) is a free program that essentially opens a Linux window within Windows. Starting Cygwin from the Windows start menu opens up a terminal window within which users type commands to compile and run Swarm models (fig. 50). Therefore, users of InSTREAM need to know a few Unix/Linux

Figure 50—The Cygwin terminal window within which Swarm models are compiled and executed.

commands (table 28). However, tasks such as moving files, creating directories, and editing files can still be done at any time using Windows programs such as Windows Explorer and Notepad.

Table 28—Basic Linux/Cygwin commands

Command[a]	Meaning
cd C:/SwarmApps-2.2	Change directories to C:/SwarmApps-2.2
cd inSTREAM_V4	Move down to subdirectory inSTREAM_V4
cd ..	Move up to the next higher directory
cd ../inSTREAM_V4.2	Move to a parallel directory inSTREAM_V4.2
cd ../..	Move two directory levels up
./instream.exe	Execute the program "instream.exe"
./instream.exe &	Execute the program in background, so the user can continue to use the terminal while the program runs.
cp Model.Setup ../SiteTwo	Create a new copy of the file Model.Setup in the directory called SiteTwo
cp Model.Setup Model.Setup.old	Create a new copy of the file Model.Setup in the current directory
cp DataDirectory ../SiteTwo -r	Create a new copy of the directory DataDirectory under the directory called SiteTwo (the option "-r" means to copy the directory and its files and subdirectories)
ls	List the files in the current directory
ls *.m	List files ending in ".m"
ls -la	List files providing details (file size, creation date, etc.)
mv Model.Setup ../SiteTwo	Move the file Model.Setup to the directory called SiteTwo
mv LiveFish.Out LiveFish_Run1.Out	Rename the file from LiveFish.Out to LiveFish_Run1.Out
mkdir SiteFiles	Make a new subdirectory called SiteFiles
rmdir SiteFiles	Remove a subdirectory called SiteFiles
grep mortFishTerrPred *.m	Search all the files ending in ".m" for the text "mortFishTerrPred"
tar xzvf AnArchiveFile.tgz	Uncompress a TAR-format archive of files. The unarchived files will be placed in the current directory
tar czvf ModelOutput.tgz results/*	Create a TAR format archive of all the files in the subdirectory called "results"; the archive is named ModelOutput.tgz
tar czvf AnArchiveFile.tgz SiteFiles	Create a TAR archive of the subdirectory SiteFiles and all its contents (including further subdirectories); when unarchived, the subdirectory will be re-created

[a] Note that in Cygwin and Unix, directories are delineated with "/", not "\" as in Windows. Directory and file names should not include spaces, and are case sensitive.

Two additional features of Cygwin that users should be aware of:

- When entering commands, you can type the first few letters of a file or directory name, then hit the tab key and Cygwin will try to complete the name. For example, the command to execute InSTREAM is `./instream.exe` (section 19.4); the user can type `./ins` and then hit tab; the computer will then complete the command (unless there is another file in the directory that also starts with "ins", in which case Cygwin cannot determine which file is desired until more of it is entered).

- You can pause a model run by entering `control-s`, and then resume its execution by entering `control-g`. You can kill a run with `control-x`. (None of these commands work if the model is running in "background.")

19.2. Installing Cygwin and Swarm

Starting with Swarm version 2.2, Swarm is installed directly from the Internet using a setup program and procedures found at http://www.swarm.org (see the "stable release" page at this site). The setup program downloads files from the Internet and installs them, so a high-speed Internet connection is required. The user actually installs a special version of Cygwin that includes Swarm as one of its libraries, so Cygwin, not Swarm, appears on the Windows "start" menu and (optionally) as a desktop icon.

One of the few options during Cygwin installation is whether it should use DOS or Unix file formats (explained in section 20.1.1). The DOS format is strongly recommended, to avoid some potentially very frustrating and obscure problems.

After installing Cygwin, you will need to set the "SWARMHOME" environment variable—an operating system variable that tells Cygwin where the Swarm libraries are when you compile InSTREAM. Follow the Swarm installation guidance to set SWARMHOME.

By default, Cygwin is installed in a directory tree starting with C:\Cygwin. (The Swarm libraries start at C:\Cygwin\usr\include\swarm). Swarm models such as InSTREAM can be stored anywhere on the same computer; by convention this document assumes that users create a new directory tree for Swarm 2.2 applications, calling it C:\SwarmApps-2.2.

When Cygwin is started, it begins in what Cygwin considers the user's home directory—usually, the user's directory under C:\Documents and Settings—**not**, as many of us would expect, at the root directory C:/.

19.3. Installing and Compiling InSTREAM

The InSTREAM software is typically distributed as a compressed archive file in TAR Gzip (.tgz) format. Installing it is a matter of uncompressing the source code files and compiling them into an executable (.exe) file. Here it is assumed that the software is distributed in a file called inSTREAMV4.2.tgz **(the actual file name is subject to change)**. The user is assumed to have created the directory c:\SwarmApps-2.2. The installation steps are:

- Copy the archive file inSTREAMV4.2.tgz into the Swarm applications directory.

- In the Cygwin terminal, uncompress the distribution file. After opening the terminal window, change directories to the Swarm applications directory by typing: `cd C:/SwarmApps-2.2`. (Use a forward slash "/", not "\".) Next, uncompress the distribution file by typing: `tar xzvf inSTREAMV4.2.tgz`. This step creates a new subdirectory (e.g., C:/SwarmApps-2.2/inSTREAM4.2/) that contains the source code and some subdirectories with example data. The source code is always distributed in one directory, with input file sets in subdirectories.

- When you look at the new directory C:/SwarmApps-2.2/inSTREAM4.2/ you will see the source code: a pair of files for each of the software's many "classes." These are the "interface" file (which declares the class's variables and methods and ends in .h) and the "implementation" file (which contains the actual program code and ends in .m). For example, the code for what model trout do is in the files "Trout.h" and "Trout.m."

- Compile the code by entering the command `make`. This step runs the compiler and linker, producing some warning statements but (unless something is wrong) successfully producing the executable version of the code, instream.exe. This step also creates an "object file" (ending in .o) for each class.

- Move the executable file instream.exe down into the subdirectory containing input files you want to use. For example, if there is a subdirectory called "ExampleRun," enter `mv instream.exe ExampleRun`. (You can also move instream.exe using Windows Explorer, etc.)

Remember to check the troubleshooting guide (section 26) if you have problems in these steps.

Many users will need to occasionally recompile the software after making such changes (discussed below) as adding different species or turning on optional output files. This can usually be done by simply repeating the `make` command: the computer figures out which source code files have been altered, recompiles them, and builds a new executable file. However, whenever an interface (.h) file is altered, it is safest to force the computer to recompile all the code. This is done by first entering the command `make clean` (which erases all the object files and the executable file), then `make`. Whenever unexpected or inexplicable things happen after recompiling with `make`, it is best to try using `make clean` and then `make` again.

19.4. Running InSTREAM: Graphics and Batch Modes

Running InSTREAM is a matter of executing instream.exe in a directory that contains the desired input files. A copy of **instream.exe must be placed in the same directory as the input files** it will use, and output files are written into the same directory. Therefore, it is usually best to create a separate subdirectory for each simulation experiment, with its input and output files and its own copy of instream. exe.

The model can be run in any directory on the computer, but **will not run on a computer that it was not compiled on**. The model can only be run within the Cygwin terminal window, not (for example) by clicking on instream.exe from Windows Explorer.

To run the model, first start the Cygwin terminal window. Then, in the window, change directories to where the model run (all the input files) resides, e.g.:

```
cd C:/SwarmApps-2.2/inSTREAM4.2/Experiment1/
```

Then the model can be started by typing the name of the executable file, preceded by "./": `./instream.exe`

(The `./` is required to tell Cygwin to look in the current directory for instream. exe. Forgetting this is a common source of frustration for new users.)

Executing the model using `./instream.exe` starts it in graphics mode, so all graphical interfaces are activated. This mode (described in section 22) is very important for demonstrating and testing the model, and it is strongly recommended that at least a trial run using graphics mode be executed any time substantial changes are made to the model or its input.

However, there are several disadvantages to graphics mode: it slows down execution of the model, it clutters up the user's computer screen, and it requires users to manually start each model run when automated experiments (section 25) are used. "Batch mode" runs InSTREAM without graphics and executes multiple-run experiments without interruption. Batch mode is invoked by starting the model with the "-b" flag: `./instream.exe -b`

When batch mode is used, a model run can be killed by typing `cntl-c` (while holding down the "control" key, enter a small c) in the terminal window. Similarly, execution can be temporarily suspended using `cntl-s` and re-started with `cntl-g`.

Swarm models can also be run in background mode so they are unaffected by the terminal window. The "&" flag invokes background mode: `./instream.exe -b &`

19.5. Known Problems and Limitations

Swarm and InSTREAM software is generally robust and reliable, but there are two known problems that "extreme users" may encounter when executing large jobs (simulating many years, many trout and many habitat cells, or executing many model runs in a single Experiment Manager job).

When running multiple simulations with the Experiment Manager (section 25), InSTREAM does not free all the memory it used during a model run when the run finishes. Consequently, memory use increases as more runs are executed; this can gradually slow execution and even bring it to a near halt as the available random access memory (RAM) is used up. This problem actually results from how Swarm was designed. Versions of InSTREAM released March 2005 or after use much less memory and free a much greater percentage of memory at the end of a run, so this problem should not affect most users. The only solution to this problem is to install more RAM.

The second problem appears to be caused by a limit that Cygwin places on the total memory allocated by a single process. Symptoms of this problem are that the model crashes for no apparent reason during a very long run, with a Swarm error message concerning the method "xmalloc," and that the crash recurs at exactly the same point if the model is restarted. Solutions to this problem are (1) searching the Cygwin documentation at http://www.cygwin.com for "Changing Cygwin's Maximum Memory" and trying the solutions found there and (2) running the model in Linux instead of Windows.

20. Input: Setup, Parameter, and Data Files

This section provides the detailed format for all the files needed to define an InSTREAM run. **Setup** files provide run-control information, **parameter** files provide equation coefficients for trout and habitat, and **data** files provide the numbers needed to define habitat cell characteristics, the initial trout populations, and the time series of flow, temperature, and turbidity that drive the model.

20.1. Common Characteristics of Input Files

All of the setup, parameter, and data input files are in ASCII. They can be maintained and edited using ASCII editors (e.g., Notepad, emacs, gvim), or by using word processor or spreadsheet software and saving them as ASCII (plain text). Spreadsheets are especially useful for maintaining most of the input files.

Because InSTREAM runs in a Unix-like environment (even under Windows), it uses **case sensitive** file names and variable names. The model needs a file named "Model.Setup" and it will not find and use files named "model.setup" or "Model. setup." Similarly, a variable named "fishParam" is different from one named "fishparam." Failing to notice case differences in file and variable names is a common source of frustration.

None of the files require values to be in any particular columns: blanks in the file are ignored. Values on the same line can be separated by one or more space or tab characters.

Many of the input files start with three header lines that are ignored by the software; these can be used to document the file type and where its information came from, and to provide column labels for the remaining lines. These header lines can be up to 200 characters long.

20.1.1. Translating files between Unix and Windows—

Transferring files between Windows and Unix-based (including Linux) operating systems can result in subtle problems because the two operating systems use different codes for the line ends in ASCII files. Hence, files created in Windows may not work in Linux, and files created in Linux may work in some, but not all, Windows programs. Cygwin can be installed so it uses either Windows (DOS) or Unix formatting. Attempting to run InSTREAM in Unix—or in Cygwin if Cygwin was installed with the Unix file format option—with files created (or edited) in Windows is likely to fail because of this problem (and Unix will not clearly indicate what the problem is).

Unix and Linux operating systems, and Cygwin, include programs "dos2unix" and "unix2dos" to convert between formats. For example, to convert a directory of files prepared in Windows for use on a Linux computer, in Linux or Cygwin type `dos2unix *` within the directory. Be aware, though, that **dos2unix and unix-2dos destroy any executable file** they attempt to convert, so be careful using them on directories that contain instream.exe.

20.1.2. Files read by the Swarm object loader—
The setup files and parameter files are read by Swarm's "object loader" facility. The object loader is a simple tool for reading in variable values for an object. Except as noted below, all the setup and parameter files must be in the following object loader format.

- The first line has only the text `@begin` (unless the first lines are comments; see below).
- There is one line per variable, with each line containing the variable name followed by its value.
- Variables need not be in any particular order.
- The variable must be spelled **exactly** as it is in the code, including upper/lower case.
- For text variables, the value text does **not** use quotation marks.
- Variables containing text **must not have trailing blanks after the text input**. For example, a variable that has values of either `NO` or `YES` must not have a blank space after the value. (The blank will be read as part of the variable's value, which makes the code unable to interpret the value. If the variable is a file name, the code will be unable to open the file because it will look for a file with a blank at the end of its name. This is another potential source of frustration.)
- Integer values should not have a decimal point.
- The last line has only the text `@end`.
- Comment lines (ignored by the computer) can be included in the file; they start with the character #. (In a few strange cases, Swarm was unable to distinguish comments, so they had to be removed.)

Example object loader files are provided in the following sections.

20.2. Setup Files

Five setup files provide run-time control information about graphical outputs, species, habitat reaches, automated experiments, and the model run itself.

20.2.1. Observer setup—

This file controls the observer swarm, which provides the graphical interfaces described in section 22. It must be named "Observer.Setup". An example is:

```
@begin

rasterColorVariable        depth
rasterZoomFactor              1
rasterResolution             40
rasterResolutionX            20
rasterResolutionY            20
takeRasterPictures           NO

@end
```

The variables in Observer.Setup are explained in table 29.

Table 29—Contents of Observer.Setup

Variable name and type	Definition
rasterColorVariable	Selects the habitat variable used to color-code habitat cells on the animation window. Valid values are depth and velocity.
rasterZoomFactor (integer)	Controls the size of the animation window. A value of 2, for example, doubles the size of the window and its contents.
rasterResolution (integer)	Controls the pixel size of the raster window. Increasing its value makes the animation sharper but uses slightly more memory. Values below about 40 make it difficult to distinguish individual fish.
rasterResolutionX (integer) rasterResolutionY (integer)	Controls the size of raster objects in the X (left-right; downstream-upstream) and Y (top-bottom; left bank-right bank) dimensions. An object's actual location (in cm) is divided by rasterResolutionX and rasterResolutionY to place it on the raster window. Smaller values of these parameters make the raster display larger. The values of rasterResolutionX and rasterResolutionY can be varied separately to exaggerate one dimension.
takeRasterPictures (text)	Determines whether the raster window is captured in files for post-processing into a movie of the simulation (see section 22.5.2). Valid values are no and yes (also NO and YES). Raster pictures should not be turned on unnecessarily; they severely reduce execution speed and generate many large output files.

20.2.2. Species setup—

The species setup file tells InSTREAM which species are being simulated and where to find the input files for each. (This file does **not**, by itself, determine how many species are modeled; see section 21.1.) The file must be called "Species. Setup." The contents of this file must match the code that defines species in the model, as explained in section 21.1.

The species setup file does **not** use the Swarm object loader format. A Species.Setup example for a two-species model is:

```
Species.Setup file, rainbow and brown trout model.

For each species, provide species class name, parameter
file name, initial population file name, and raster
display color.

Rainbow
Rainbow.Params
RbTInitialPops.Data
OrangeRed

Brown
Brown.Params
BTInitialPops.Data
brown
```

The species setup file starts with a block of three comment lines that are ignored by the software, followed by a blank line. Then come a block of lines for each species in the model. Each such block includes:

- A line with the species name, which must exactly match the name of the source code class for the species. (If this file contains a species "DollyVarden," then the source code files DollyVarden.h and DollyVarden.m must have been compiled when instream.exe was created.)
- A line with the name of the parameter file that the species uses. (One parameter file can be used by more than one species.)
- A line with the name of the initial population data file for the species.
- A line containing the name of the color used for the species in the animation window. Allowable values include all the common colors (red, green, blue, etc.) plus a large number of exotic color names. (A complete list of colors can be obtained in Unix-like systems—but not Cygwin—via the command `showrgb`.) Invalid color names result in white being used.
- A blank line, if another species follows.

20.2.3. Reach setup—

This setup file, which must be named "Reach.Setup," specifies the number of habitat reaches, how reaches are linked, and what input files should be used for each reach. The file does **not** use the object loader format. The file contains:

- Three header lines that are ignored by the computer.
- A blank line.
- A block of lines for each reach. These blocks start with a line containing only the word REACHBEGIN and end with a line containing only the word REACHEND. There can be multiple blank lines between these blocks.

The block of lines for each reach contains a separate line for each of the reach's setup variables. The line contains the variable name, one or more spaces, then the variable value. These lines need not be in any particular order within the block. An example Reach.Setup file follows, with variables explained in table 30.

```
Reach.Setup file.
Provide one block for each reach.
Example input.

REACHBEGIN
reachName                          WeejakTrib
habParamFile                       LJCHab.Params

habDownstreamJunctionNumber        2
habUpstreamJunctionNumber          3

cellDataFile                       LJCWeejCell.Data
flowFile                           WeejTestFlow.Data
temperatureFile                    LJCLowTemp.Data
turbidityFile                      LJCLowTurbidity.Data

barrierX                           22.5
barrierX                           33.5

hydraulicsFile1                    LJCWjHyd1.Data
hydraulicsFile2                    LJCWjHyd2.Data
hydraulicsFile3                    LJCWjHyd3.Data
hydraulicsFile4                    LJCWjHyd4.Data

REACHEND

REACHBEGIN
reachName                          LowerMainstem
habParamFile                       LJCHab.Params

habDownstreamJunctionNumber        4
habUpstreamJunctionNumber          2
```

(etc. for remaining reaches)

Table 30—Contents of the Reach.Setup file

Reach variable and type	Definition
reachName (text)	The user-defined name for a reach (up to 30 characters).
habParamFile (text)	The name of the reach's habitat parameter file.
habDownstreamJunctionNumber (integer)	The junction number for the reach's downstream end. (Chapter 2 explains junction numbers.)
habUpstreamJunctionNumber (integer)	The junction number for the reach's upstream end.
cellDataFile (text)	The name of the reach's cell data file. The file name can be up to 35 characters long.
flowFile (text) temperatureFile turbidityFile	The names of the flow, temperature, and turbidity data files for the reach to use. (Different reaches can use the same files.) These file names can be up to 35 characters long.
barrierX (float)	Optional values that define the location of barriers (which fish can pass downstream but not upstream). One barrierX line is provided for each barrier. Its value is the location of the barrier, as distance (m) from the downstream end of the reach.
hydraulicsFile1 (text) hydraulicsFile2 hydraulicsFile3 hydraulicsFile4 hydraulicsFile5	The names of up to five hydraulic data files; the RHABSIM files that each provide a lookup table of cell depth and velocity as a function of flow (explained in chapter 3). (If fewer than five hydraulic data files are used, then fewer than five lines are used.) Each file provides hydraulic data for a unique range of flows. File names must be ordered from lowest to highest range of flows: hydraulicsFile1 must cover the lowest range of flows, and the last hydraulic data file must cover the highest range of flows.

20.2.4. Model setup—

The model setup file must be called "Model.Setup". It contains basic variables controlling a model run, as illustrated in this example. The file contents are explained in table 31.

```
# Model Setup File for Little Jones Creek Cutthroat Trout Model
# Test runs 2/04

@begin

randGenSeed            32461
numberOfSpecies        1

runStartDate           10/1/1990
runEndDate             9/30/2001
fishOutputFile         LiveFish.out
fishMortalityFile      DeadFish.out
reddMortalityFile      ReddMortality.out
reddOutputFile         Redds.out
popInitDate            10/1/2000
```

```
fileOutputFrequency    10
appendFiles            YES

shuffleYears           0
shuffleYearReplace     0
shuffleYearSeed        737899

tagFishColor           tomato

siteLatitude           42

@end
```

Table 31—Contents of the Model.Setup file

Reach variable and type	Definition
randGenSeed (integer)	The seed value for the random number generator used for all stochastic processes except year shuffling. It can be any positive integer.
numberOfSpecies (integer)	The number of species to simulate. The number should not exceed the number of species defined in the species setup file. If three species are defined in Species.Setup but numberOfSpecies is two, then the first two species in Species.Setup are simulated. No more than 10 species can be modeled.
runStartDate runEndDate (date: mm/dd/yyyy)	The dates simulations begin and end.
fishOutputFile (text)	The name of the output file for statistics on live fish.
fishMortalityFile (text)	The name of the output file for statistics on fish mortality.
reddOutputFile (text)	The name of the output file for statistics on redds.
reddMortalityFile (text)	The name of the output file for statistics on egg mortality.
popInitDate (date)	The date used to identify initial trout population data from the initial population data file. This date need not be the same as runStartDate.
fileOutputFrequency (integer)	The frequency with which file output is written. If set to 1, output is written for each simulated day; if set (for example) to 10, output is written only each 10th day. (This output is the model's state on the date when output is written, not an average for the period between output dates.)
appendFiles (boolean: yes or no)	Whether existing output files should be appended (instead of over-written) at the start of each model run. Valid values are 1 for yes and 0 for no.
shuffleYears (boolean)	Whether simulation years should be randomly shuffled (explained in chapter 2). Valid values are 1 for yes and 0 for no.
shuffleYearReplace (boolean)	Whether, if simulation years are shuffled, the randomization is with replacement (so years can be used more than once). Valid values are 1 for yes and 0 for no.
shuffleYearSeed (integer)	The seed for the random number generator used only for year shuffling. It can be any positive integer.
tagFishColor (text)	The color that fish turn, on the animation window, if "tagged."
siteLatitude (floating point)	The latitude of the study site, in degrees north (used to calculate day lengths).

20.2.5. Experiment setup—

The experiment setup file "Experiment.Setup" controls InSTREAM's experiment manager. This feature is complex enough that section 25 is dedicated to it.

Users are reminded to inspect Experiment.Setup any time InSTREAM appears to be ignoring a setup variable or parameter (yet another common source of frustration!).

20.3. Parameter Files

InSTREAM requires two kinds of parameter files, for habitat and trout. These files provide the values for all the parameters defined in chapter 2. Each habitat reach, and each trout species, must have a parameter file assigned to it (this assignment is done in the reach and species setup files). More than one reach, or species, can use the same parameter file. Parameter files use the Swarm object loader format (section 20.1.2).

20.3.1. Habitat parameter file—

The name of the habitat parameter file for a reach is defined by the user in the reach's setup file (section 20.2.3). By convention, the file name includes the reach name, the syllable "Hab," and the extension ".Params"; for example, "SmithCrkMiddleReachHab.Params."

Habitat parameter files must include exactly the same variables as in the following example; these parameters are defined in chapter 2 of this report. However, the order in which parameters appear does not matter.

```
@begin

habSearchProd              7.0E-7
habDriftConc               1.50E-10
habDriftRegenDist             500
habPreyEnergyDensity         2500
habMaxSpawnFlow               4.0
habShearParamA              0.019
habShearParamB              0.383
habShelterSpeedFrac           0.3

@end
```

20.3.2. Trout parameter file—

The user defines the name of the parameter file for each trout species in the species setup file (section 20.2.2). By convention, the file name includes the species name, perhaps the site name, the syllable "Trout," and the extension ".Params"; an example is "SmithCrkBrownTrout.Params."

All trout parameter files must include exactly the same variables as in the following example; the parameters are defined in chapter 2. The software checks to make sure all the trout parameters are initialized: if any trout parameters are missing from this file, an error statement will be issued and execution will stop.

```
@begin

fishCaptureParam1            1.6
fishCaptureParam9            0.5

fishCmaxParamA               0.628
fishCmaxParamB               -0.3
fishCmaxTempF1               0.05
fishCmaxTempF2               0.05
fishCmaxTempF3               0.5
fishCmaxTempF4               1
fishCmaxTempF5               0.8
fishCmaxTempF6               0
fishCmaxTempF7               0
fishCmaxTempT1               0
fishCmaxTempT2               2
fishCmaxTempT3               10
fishCmaxTempT4               22
fishCmaxTempT5               23
fishCmaxTempT6               25
fishCmaxTempT7               100

fishDetectDistParamA         4.0
fishDetectDistParamB         2.0

fishEMForUnknownCells        0.1

fishEnergyDensity            5900

fishFecundParamA             0.11
fishFecundParamB             2.54

fishFitnessHorizon           90

fishMaxSwimParamA            2.8
fishMaxSwimParamB            21
fishMaxSwimParamC            -0.0029
fishMaxSwimParamD            0.084
fishMaxSwimParamE            0.37

fishMinFeedTemp              2
fishMoveDistParamA           20
fishMoveDistParamB           2
```

fishPiscivoryLength	15
fishRespParamA	30
fishRespParamB	0.784
fishRespParamC	0.0693
fishRespParamD	0.03
fishSearchArea	20000
fishSpawnEggViability	0.8
fishSpawnDSuitD1	0.0
fishSpawnDSuitD2	5.0
fishSpawnDSuitD3	50.0
fishSpawnDSuitD4	100.0
fishSpawnDSuitD5	1000.0
fishSpawnDSuitS1	0.0
fishSpawnDSuitS2	0.0
fishSpawnDSuitS3	1.0
fishSpawnDSuitS4	1.0
fishSpawnDSuitS5	0.0
fishSpawnEndDate	5/31
fishSpawnMaxFlowChange	0.2
fishSpawnMaxTemp	13
fishSpawnMinAge	1
fishSpawnMinCond	0.98
fishSpawnMinLength	12
fishSpawnMinTemp	8
fishSpawnProb	0.04
fishSpawnStartDate	4/1
fishSpawnVSuitS1	0.0
fishSpawnVSuitS2	0.0
fishSpawnVSuitS3	1.0
fishSpawnVSuitS4	1.0
fishSpawnVSuitS5	0.0
fishSpawnVSuitS6	0.0
fishSpawnVSuitV1	0.0
fishSpawnVSuitV2	10.0
fishSpawnVSuitV3	20.0
fishSpawnVSuitV4	75.0
fishSpawnVSuitV5	100.0
fishSpawnVSuitV6	1000.0
fishSpawnWtLossFraction	0.2
fishTurbidExp	−0.0711

fishTurbidMin	0.1
fishTurbidThreshold	5.0
fishWeightParamA	0.0124
fishWeightParamB	2.98
mortFishAqPredD1	20
mortFishAqPredD9	5
mortFishAqPredF1	18
mortFishAqPredF9	0
mortFishAqPredL1	4
mortFishAqPredL9	8
mortFishAqPredMin	0.9
mortFishAqPredP1	1.00E-05
mortFishAqPredP9	2.00E-06
mortFishAqPredT1	6
mortFishAqPredT9	2
mortFishAqPredU1	5
mortFishAqPredU9	80
mortFishConditionK1	0.3
mortFishConditionK9	0.6
mortFishHiTT1	30
mortFishHiTT9	25.8
mortFishStrandD1	-0.3
mortFishStrandD9	0.3
mortFishTerrPredD1	5
mortFishTerrPredD9	150
mortFishTerrPredF1	18
mortFishTerrPredF9	0
mortFishTerrPredH1	500
mortFishTerrPredH9	-100
mortFishTerrPredL1	6
mortFishTerrPredL9	3
mortFishTerrPredMin	0.99
mortFishTerrPredT1	10
mortFishTerrPredT9	50
mortFishTerrPredV1	20
mortFishTerrPredV9	100
mortFishVelocityV1	1.8
mortFishVelocityV9	1.4
mortReddDewaterSurv	0.9
mortReddHiTT1	30

```
mortReddHiTT9                    21
mortReddLoTT1                    -3
mortReddLoTT9                     0
mortReddScourDepth                5

reddDevelParamA          -0.000253
reddDevelParamB            0.00134
reddDevelParamC           3.21E-05
reddNewLengthMean              2.8
reddNewLengthStdDev            0.2
reddSize                      1200

@end
```

20.4. Data Files

Four kinds of files are used to define habitat and the initial trout population.

20.4.1. Cell data—

There is one cell data file for each reach modeled. The name of a reach's cell data file is specified by the user in the reach setup file. The convention is for these file names to include the reach name and end in "Cell.Data" (e.g., SmithCrkLower-ReachCell.Data).

The cell data files have a unique format because they must include a number of variables for each cell, including the cell boundaries. Cells occur in rows across the channel called transects. All cells on a transect have the same coordinates in the X (upstream-downstream) dimension. Cell boundaries are defined by input identifying (a) the X coordinates of the boundaries between transects, and (b) the Y (across-channel) coordinates of the boundaries between cells on each transect. (These conventions for defining cell boundaries are explained in chapter 2.)

The X coordinates and transect numbers increase from downstream to upstream, and Y coordinates and cell numbers increase from the right to left banks when facing downstream. The X coordinate is zero at the downstream end of the model reach, and Y is zero at the right end of each transect. (The first X coordinates in the cell data file are greater than zero because they are the distance the first transect extends upstream from zero. Likewise, the Y coordinates are all greater than zero because they are the distance of a cell's left side from zero; for the first cell on a transect, the Y coordinate will also be the cell's width.)

All distances in this file are in meters and converted inside the model code to cm, which are used for all computations. (This is a convenient violation of the units conventions established in chapter 1.)

The file starts with three rows of header information that are ignored by the software. The third header line usually provides the column headings for the rest of the file.

Starting on the fourth line, the file provides one line of information for each cell, with the data for each cell in columns. No particular spacing is required as long as there are one or more spaces or tabs between values. **None of the values can be left blank**. The following columns are in this file.

- The transect number. Transects are numbered using integers in ascending order, starting with 1 for the most downstream transect. The transect number is the same for all the data file rows representing cells on the same transect.

- The cell number. These are integers in ascending order, starting with 1, and start over for each transect. (Cell 1 is the first cell on the right bank of each transect, facing downstream.)

- The X coordinate of the **upstream** end of the cell. This upstream X coordinate is the same for all cells on the same transect. These coordinates are the longitudinal distance between (a) the downstream end of the cells on the first transect (where X is zero) and (b) the upstream end of the cells on the current transect. For the first transect, this upstream X coordinate equals the cells' length in the upstream-downstream direction.

- The Y coordinate of the left boundary of the cell (facing downstream). This left Y coordinate increases with cell number across each transect.

- The proportion of the cell (0 to 1) having velocity shelters for drift-feeding.

- The distance to hiding cover for the cell, in meters.

- The proportion of the cell with suitable spawning gravel.

Here is a (partial) example cell data file.

```
Cell data for Little Jones Creek, Lower site
Test file by SFR, 4/1/99
```

Transect No.	Cell No.	UpstreamX	LeftY	FracShelter	DistToHide	FracSpawn
1	1	5.0	4	0	2.5	0.1
1	2	5.0	5.2	0.5	0.6	0.3
1	3	5.0	6.1	0.2	0.4	0.9
1	4	5.0	8.3	0.3	0.5	0.8
1	5	5.0	11	0.2	1.5	0.2
1	6	5.0	13.6	0.5	0.3	0.0
1	7	5.0	15.8	0.6	0.3	0.2
2	1	13.6	4.6	0.1	5	0.1
2	2	13.6	7.2	0.3	3	0.9
2	3	13.6	8.4	0.3	2	1.0
2	4	13.6	10.7	0.3	2	0.9
2	5	13.6	13	0.1	1.5	0.2

2	6	13.6	15.1	0.7	0.3	0.4
2	7	13.6	16.6	0.7	0.3	0.7
2	8	13.6	17.9	0.9	0.2	0.3
3	1	26.7	0.5	0.3	5	0.0
3	2	26.7	2.2	0.2	4	0.7
3	3	26.7	4.6	0.2	5	0.1
3	4	26.7	7	0.3	5	0.5
3	5	26.7	9.6	0.1	4	0.5
3	6	26.7	11.3	0.3	2	0.5
4	1	39.4	2	0.6	5	0.1
4	2	39.4	5	0.4	5	0.3
4	3	39.4	7	0.2	3	0.3
4	4	39.4	9.2	0.2	4	0.4
4	5	39.4	11.4	0.4	4	0.5
4	6	39.4	14.2	0.6	2	0.6
4	7	39.4	15.2	0.4	0.2	0.2

. . .

20.4.2. Hydraulic data—

To model how the depth and velocity of each cell varies with flow, InSTREAM imports files produced by the RHABSIM hydraulic simulation software (explained in chapter 3). There can be as many as five of these hydraulic data files, each representing a unique range of flows. The names of these files are provided by the user for each habitat reach in the reach setup file (section 20.2.3). InSTREAM uses these files exactly as produced by RHABSIM. The naming convention for hydraulic data files is to include the reach name, the file number (from lowest to highest range of simulated flows, the order the file names appear in the reach setup file), and end with "Hyd.Data." Examples are SmithCrk1Hyd.Data, SmithCrk2Hyd.Data, etc.

20.4.3. Initial population data—

This file specifies the initial population of trout that is created at the start of a model run. One file is provided for each species, with the file name provided by the user in the species setup file. When multiple reaches are simulated, initial population data for all reaches are provided in the same file. The file name convention is to include the species and study site names, and end in "InitPops. Data" (e.g., SmithCrkRainbowInitPops.Data).

The file starts with three comment lines that are ignored by the software; the third line is usually column headings. Next come lines of data that each specify the initial number and size of trout of one age, for one reach.

The following example file initializes one species, with four ages, for two reaches. Two alternative sets of initial population characteristics are provided, each with its own initialization date in the first column.

```
Initial Fish Populations for LJC, Made up by SFR 1/30/02

For Test of multiple reaches

Init Date    Age    Number    MeanLength, cm    StdDevLength         Reach
10/1/1999     0      212          6.1               0.94         LowerMainstem
10/1/1999     1       38         11.6               1.1          LowerMainstem
10/1/1999     2        4         17.1               2.5          LowerMainstem
10/1/1999     3       10         19.9               3.2          LowerMainstem
10/1/1999     0      112          5.1               0.93         UpperMainstem
10/1/1999     1       28         10.6               1.1          UpperMainstem
10/1/1999     2        2         15.1               2.4          UpperMainstem
10/1/1999     3       10         19.9               3.1          UpperMainstem
10/1/2000     0       22          6.1               0.93         LowerMainstem
10/1/2000     1        3         11.6               1.0          LowerMainstem
10/1/2000     2        4         17.1               2.5          LowerMainstem
10/1/2000     3        0         19.9               3.1          LowerMainstem
10/1/2000     0       11          5.1               0.98         UpperMainstem
10/1/2000     1        2         10.6               1.1          UpperMainstem
10/1/2000     2        2         15.1               2.5          UpperMainstem
10/1/2000     3        0         19.9               3.2          UpperMainstem
```

Each line includes the following values:

- The initialization date. This date is used only as an index allowing the initial population file to contain more than one set of initialization data, which the user selects via the Model.Setup file variable *popInitDate*. The above example allows the user to start the model using data representing October first of either 1999 or 2000; either of these two initialization data sets can be selected in Model.Setup by setting its value of *popInitDate* to either 10/1/1999 or 10/1/2000. The initialization date does not need to be the date on which simulations actually begin. Arbitrary values of the initialization date can be used to provide alternative initial population scenarios for simulation experiments.
- The age of fish to initialize.
- The number of fish of the specified age to create.
- The mean length (cm) of the initial fish.
- The standard deviation (cm) of the initial fish. The length of each fish is drawn randomly from a normal distribution defined by the mean and standard deviation specified here.
- The name of the reach for which the fish are initialized. This name must be exactly equal to one of the reach names provided in the reach setup file.

These lines must be in the order illustrated in the above example. Lines for one initialization date must be grouped together. Within the block for each initialization date, lines for each reach must be together. Finally, lines for each reach must be sorted in increasing order by age.

There is no limit to how many ages can be initialized, and the initial ages need not correspond to the age classes used to summarize output (section 21.2).

20.4.4. Flow, temperature, and turbidity time series data—

InSTREAM uses daily input values of flow, temperature, and turbidity, for each reach. Each of these variables is provided in a separate file (each data file provides one time series of one variable). The names of the flow, temperature, and turbidity data files used by each reach is specified by the user in the reach setup file. The file names usually include the reach name, the type of data, and ".Data" (e.g., SmithCrkFlow.Data, SmithCrkTurbid.Data).

These data files are read by the EcoSwarm class TimeSeriesInputManager. The format for each is:

- The first three lines are headers ignored by the software. The third line usually provides column headings.
- Each remaining line provides the value for one day. Each line starts with the date (mm/dd/yyyy), followed by the daily value.

The data lines must be in increasing order by date, and no days (including leap days) may be omitted. Flow values are in m^3/s, temperatures in °C, and turbidity in nephelometric turbidity units (NTU). An example is:

```
Temperature data for LJC Lower Site. From 1999-00 measured values
Assembled 1/11/01
Date          Temperature
10/1/1987        11.6
10/2/1987        11.6
10/3/1987        11.5
10/4/1987        11.4
10/5/1987        11.4
10/6/1987        11.3
10/7/1987        11.3
10/8/1987        11.2
10/9/1987        11.1
10/10/1987       11
10/11/1987       10.9
10/12/1987       10.7
...
```

These files may include dates outside the range being simulated in any particular model run; InSTREAM simply finds the values it needs from within the data files. Flow data must include the day after the last simulation day (for modeling redd scouring mortality; chapter 2).

Spreadsheet software is especially convenient for building these time series data files. However, when data are imported to a spreadsheet it can convert the dates to a different format (e.g., dates show up in the spreadsheet as "10/1/87" instead of "10/1/1987," a common source of frustration). It can be necessary to re-format the date column in the spreadsheet before saving it as the ASCII file used by InSTREAM. Another frequent problem is spreadsheet columns that are slightly too narrow, so some dates (e.g., 10/15/1987 but not 1/5/1987) are saved as "#######". Parsing errors also commonly occur because the number of characters in a date changes. If the date column starts at 1/1/2001, the spreadsheet may decide that the date column need be only 8 characters wide. When dates like 10/10/2001 are reached, the last characters end up in the next column, ruining the file. Always inspect spreadsheets carefully before exporting data for input to InSTREAM.

21. Changing Species and Age Classes

Minor changes to the software's source code are needed to change either the species represented by InSTREAM or the number of age classes used to summarize and report output. These changes do not require programming skills if the following directions are followed, but it is recommended that users save their original source code separately before making the changes.

21.1. Adding or Removing Species

InSTREAM can use up to 10 species. The number of species included in a run is set in the Model.Setup file, and the files used by each species are specified in the Species.Setup file. However, these setup files can only refer to species that exist in the software as a subclass of the "Trout" class. To exist in the software, a species must have its own interface file (.h) and implementation (.m) file. Therefore, adding a species is simply a matter of creating an interface and implementation file for the species; and telling the software to include those files when it is compiled, by including the new file names in the Makefile (explained below). Likewise, species can be removed by deleting their files and removing reference to them from the Makefile. And one species can be replaced by another by simply replacing the species name in all the locations discussed here.

It is critical that the name of the new species' interface and implementation files exactly match the species' name in the Species.Setup file. The new .h and .m files must of course be in the same directory as the rest of the source code.

The following steps add a species to InSTREAM. As an example, the rare blarney trout is added to a model already containing rainbow and brown trout. If any of the six changes are incomplete, the software may exit with an error or crash without explanation.

Step 1. Create the species interface file. This is easiest done by copying and editing the .h file of an existing species. For example, the file "Rainbow.h" can be copied to a new file "Blarney.h" and edited to:

```
#import "Trout.h"
@interface Blarney : Trout
{

}
+ createBegin: aZone;

@end
```

Step 2. Create the species implementation file. Again, this is easiest done by copying and editing the .m file of an existing species. The new "Blarney.m" file should be edited to:

```
#import "globals.h"
#import "Blarney.h"

@implementation Blarney

+ createBegin: aZone
{
 return [super createBegin: aZone];
}

@end
```

Step 3. Add the new species to the TroutModelSwarm.h file. Near the top of TroutModelSwarm.h is a series of "#import" statements that say which other .h files are used by TroutModelSwarm.h. Simply add a new #import statement for the new species, just like the statement for existing species:

```
#import <stdlib.h>
#import <objectbase/Swarm.h>
#import <analysis/Averager.h>

#import "TroutModelSwarmP.h"

#import "globals.h"
#import "Rainbow.h"
#import "Brown.h"
#import "Blarney.h" // New species added here
#import "Redd.h"
#import "HabitatSpace.h"
#import "FishParams.h"
```

Step 4. Modify the Species.Setup file by adding lines for the new species (see section 20.2.2).

Step 5. Modify the Model.Setup file to increase the parameter *numberOfSpecies*. The value of this parameter must not be greater than the number of species listed in Species.Setup.

Step 6. Edit the Makefile in three places. In the source code directory is a file called "Makefile," which provides the compiler with directions for which source code files to include in InSTREAM. Users need not try to understand the Makefile, but can just add the new species wherever an existing species is found. First, the OBJECTS statement must have the new species' class file in it. This statement should look like (your makefile may not look exactly like this):

```
OBJECTS=Trout.o Cell.o Vector.o HabitatSpace.o \
    Redd.o \
    TroutModelSwarm.o \
    TroutObserverSwarm.o \
...


\
    Barrier.o \
\
    Rainbow.o \
    Brown.o \
    Blarney.o \
\
```

(The "\" characters just mean that the statement continues onto the next line.)
Second, add the new .h file to this statement:

```
TroutModelSwarm.o: TroutModelSwarm.[hm] globals.h \
Rainbow.h Brown.h Blarney.h HabitatSpace.h \
FishParams.h DEBUGFLAGS.h
```

Finally, a line must be added to the statements for each species:

```
Rainbow.o : Rainbow.[hm] DEBUGFLAGS.h
Brown.o : Brown.[hm] DEBUGFLAGS.h
Blarney.o : Blarney.[hm] DEBUGFLAGS.h
...
```

These steps add a new species that has exactly the same formulation as the other species (although the new species can have its own parameter values). Differences in formulation among species can be implemented by copying the relevant methods from Trout.h and Trout.m to the species' source code files and revising them.

After these changes are made, the software must be re-compiled to create a new version of the executable file instream.exe (the `make clean` step is recommended; section 19.3).

21.2. Changing Age Classes

The age classes used to summarize model results must be defined in the InSTREAM source code, but it is easy to modify the code to add or remove age classes. (The number of age classes do not affect simulations at all, only how statistical summaries of results are calculated and reported. Each model fish has an actual age that is not affected by these age classes.)

Most of the changes are made in the file TroutModelSwarm.m. The first is in the following lines, which appear in the method "buildObjects":

```
ageSymbolList = [List create: modelZone];

    Age0  = [Symbol create: modelZone setName: "Age0"];
    [ageSymbolList addLast: Age0];
    Age1  = [Symbol create: modelZone setName: "Age1"];
    [ageSymbolList addLast: Age1];
    Age2  = [Symbol create: modelZone setName: "Age2"];
    [ageSymbolList addLast: Age2];
    Age3Plus = [Symbol create: modelZone setName: "Age3Plus"];
    [ageSymbolList addLast: Age3Plus];
```

These lines define the age classes, and can be edited to change the number of classes. For example, if a study site includes many large, older fish, it could be desirable to use two more classes, by revising this code to:

```
ageSymbolList = [List create: modelZone];

  Age0  = [Symbol create: modelZone setName: "Age0"];
    [ageSymbolList addLast: Age0];
  Age1  = [Symbol create: modelZone setName: "Age1"];
    [ageSymbolList addLast: Age1];
  Age2  = [Symbol create: modelZone setName: "Age2"];
    [ageSymbolList addLast: Age2];
  Age3  = [Symbol create: modelZone setName: "Age3"];
    [ageSymbolList addLast: Age3];
  Age4 = [Symbol create: modelZone setName: "Age4"];
  [ageSymbolList addLast: Age4];
  Age5Plus = [Symbol create: modelZone setName: "Age5Plus"];
  [ageSymbolList addLast: Age5Plus];
```

The second change must be made in the TroutModelSwarm method "getAgeSymbolForAge", which assigns an age class to a fish, using the fish's age. For the default four age classes, this method is:

```
-  (id <Symbol>) getAgeSymbolForAge: (int) anAge
{
   int fishAge = anAge;
   if(fishAge >= 3)
{
   fishAge = 3;
}

   return [ageSymbolList atOffset: fishAge];
}
```

For the above example in which two more age classes are added (so age 5 and older fish are combined), the method must be changed to:

```
-  (id <Symbol>) getAgeSymbolForAge: (int) anAge
{
   int fishAge = anAge;
    if(fishAge >= 5)
{
   fishAge = 5;
}

   return [ageSymbolList atOffset: fishAge];
}
```

(In this code, `fishAge` is a temporary local variable, not the fish's actual age.)

Finally, the file TroutModelSwarm.h must be edited to declare the new age class symbols. Change this block of code:

```
id <Symbol> Age0;
id <Symbol> Age1;
id <Symbol> Age2;
id <Symbol> Age3Plus;
```

to this:

```
id <Symbol> Age0;
id <Symbol> Age1;
id <Symbol> Age2;
id <Symbol> Age3;
id <Symbol> Age4;
id <Symbol> Age5Plus;
```

After these changes are made and the code re-compiled (using `make clean` first), InSTREAM will automatically use the new age classes to report results that are broken out by age class.

22. Graphical Interfaces

When the InSTREAM software is executed in graphics mode (section 19.4), a number of graphical interfaces are provided. Most of these are simply graphs that display results, but the animation window is truly an interactive interface to the model, allowing users to probe deeply to observe and even alter the state of individual habitat cells, fish, and redds. Control panels can be used to start, stop, or step through model runs.

Users just interested in starting up their first model runs only need to know that they must hit the "Start" button on the first control panel twice to get execution underway.

It is important to understand that the graphical interfaces are updated only at the end of each daily time step, after all other scheduled actions are complete (the schedule is defined in chapter 2). Mouse clicks and other input to the interfaces are accepted only at the end of the time step: you cannot stop a simulation part way through a day's schedule (but you can step through a day's actions; see below), and displayed information reflects the model's state at the end of a time step.

When the model is started in batch mode (section 19.4), none of these graphics appear. Instead, simulations start immediately and run until completed (or until something goes wrong, or the user kills the job).

22.1. Main Control Panel

As soon as InSTREAM is started in graphics mode (by typing ./instream.exe in the Cygwin terminal window), the main control panel opens up (fig. 51). This panel essentially operates the Experiment Manager, which then starts the model runs. The Experiment Manager (section 25) may be set up so only one model run is executed, or so that a multirun experiment is executed. Users need to know only the following:

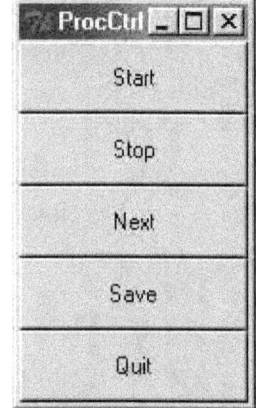

Figure 51—Main control panel.

- Model execution is started by hitting the "Start" button once (which creates the Experiment Manager), then a second time (which starts the first model run). After the "Start" button is hit the second time, control of the first model run is passed to the Model Run Controller (described below).

- If the Experiment Manager is set up for a single run, execution can be terminated cleanly by hitting the "Quit" button after the run is finished. If the Experiment Manager is set up for multiple runs, the "Start" button must be hit to start each run. After each run finishs, nothing happens until this button is hit again.

- The Save button is supposed to tell Swarm to preserve the spatial arrangement of graphical interface windows on the screen and use it next time the code is executed. However, this facility is not reliable and the hidden file it creates can become corrupted and cause frustrating problems. It is recommended that users not use it.

- The Quit button stops execution and closes the code.

22.2. Model Run Controller

When a model run has been started from the main control panel, it opens a second control panel labeled "Model Run Controller" (fig. 52). This panel displays the simulation's current time step (the number of simulation days that have been completed, starting with zero). It also has six buttons that can be used to control execution.

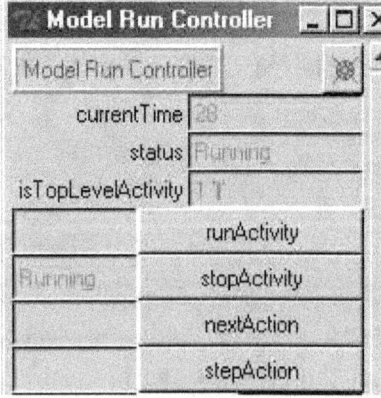

Figure 52—Model run controller.

- "runActivity" causes the model to resume execution after it has been paused.
- "stopActivity" causes the model to pause execution (after the current time step has finished).
- "nextAction" causes execution of one more time step, after which execution is again stopped. (Users may have to hit this button a number of times before it executes a full time step; it seems to work well after the "stepUntil" button is used, also.)
- "stepAction" causes execution of one individual action in the model's schedule (e.g., spawn, move, grow, die).
- "stepUntil" causes execution to continue until it reaches (but does not yet execute) a specified time step. To the right of this button is a window where the time step is input. For example, if the model is stopped at time step 5 and the user wants it to continue for 3 more days, the user would enter "9" in the stepUntil input window (by typing "9" **then hitting the Enter key**), and press stepUntil.
- "terminate" causes execution to end the current model run, passing control back to the main control panel. The proper way to shut down execution during a model run is to hit this button, then "Quit" on the main control panel.

22.3. Habitat Probe Displays

A "probe display" window (probe displays are explained in section 22.5) is opened for each habitat reach. These are small windows (fig. 53) displaying key habitat variables: the current date, flow, temperature, turbidity, etc. The first variable displayed is the reach's name.

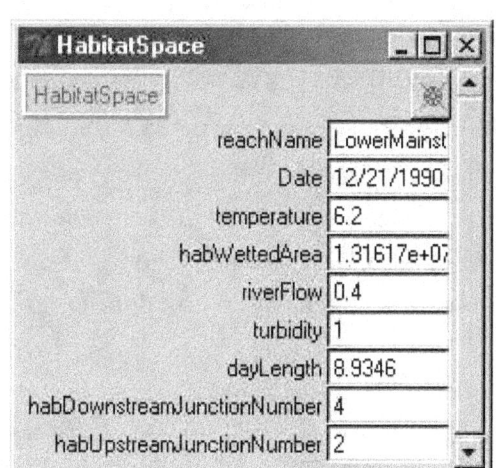

Figure 53—Habitat probe display. One of these windows is opened for each habitat reach.

22.4. Graphs

Several graphs provide summary information on the trout population. These graphs report the status of **all** trout in the model; they are not broken out by habitat reach, species, age, or any other characteristic.

- A line graph (time series plot) of the cumulative number of trout that have died of each mortality source (fig. 54). The X axis is the number of simulation days. Clicking on any mortality source in the graph's legend highlights the line for that kind of mortality.
- A bar graph showing how many fish are currently alive, by age. (This graph reports actual age of all fish, in years; it does not use the age classes used for file output.)
- Histograms showing the number of fish versus cell depth and velocities. These histograms show the number of fish occupying cells in each depth (or velocity) category. (They do not by themselves represent "preference" for depth or velocity because they do not account for how much cell area there is of each depth and velocity.) The number and size of histogram bins is set in the source code file TroutObserverSwarm.m, in the method "buildObjects" (look for the line "velocityHisto = [EZBin createBegin: obsZone]"). By default, both the depth and velocity histograms have 10 bins, each representing 10 cm of depth or 10 cm/s of velocity, with values above 100 reported as outliers.

22.5. Animation Window

The primary interactive interface is the animation window, which shows the habitat cells, fish, and redds as the simulation proceeds (fig. 55). An animation window is opened for each habitat reach. The window displays the habitat cells as a plan view (looking from the top down) map, with flow going from right to left. The size of the window can be adjusted in both dimensions using parameters in the observer setup file.

The cells are shaded by either depth or velocity, according to the observer setup file parameter *rasterColorVariable*. Cells are black when their depth (or velocity) is zero, and become lighter as depth (or velocity) increases. (The shading scheme is defined in the method drawSelfOn in file Cell.m.)

Fish appear as horizontal line segments in the cell that each fish occupies. Fish colors depend on species and are set in the species setup file. The length of the fish line is proportional to the fish's length. (The line length can be adjusted by changing the variable FISH_LENGTH_COEF in the source code file "globals.h"; the code must be re-compiled, using the `make clean` step, to make this change

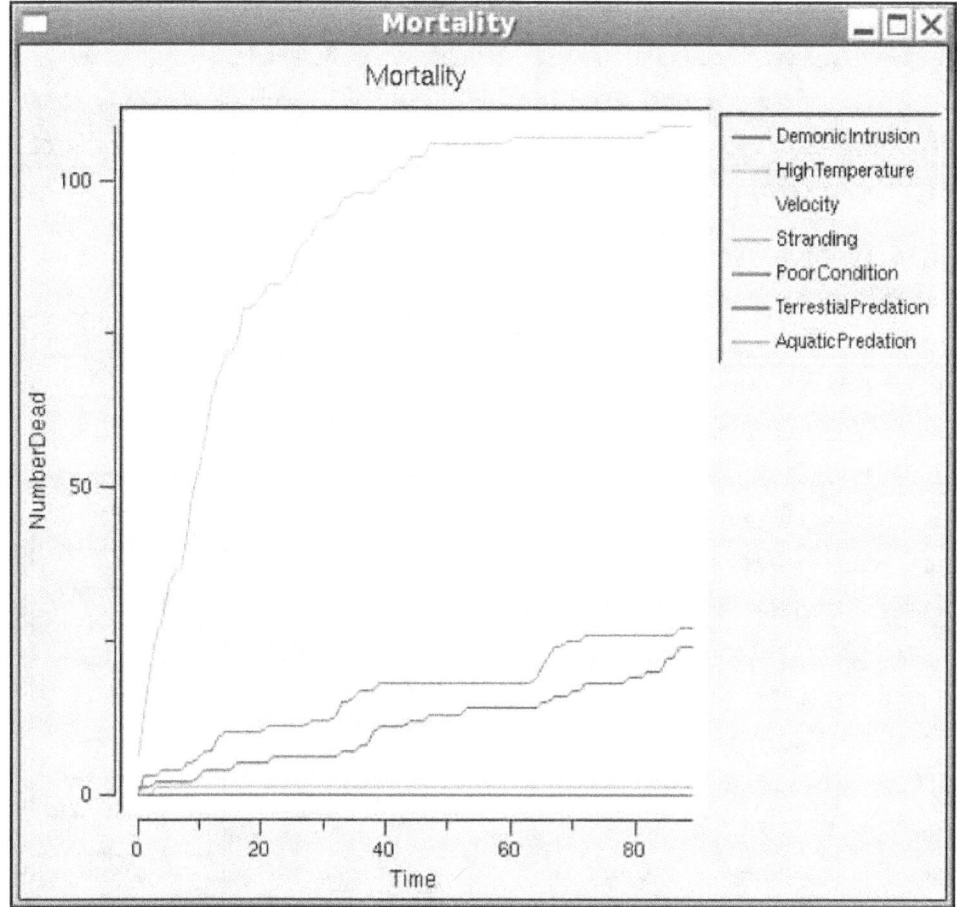

Figure 54—Graph of mortality output. "Time" on the X axis is the number of days simulated. The Y axis is the total number of trout that have died of each of the seven mortality sources.

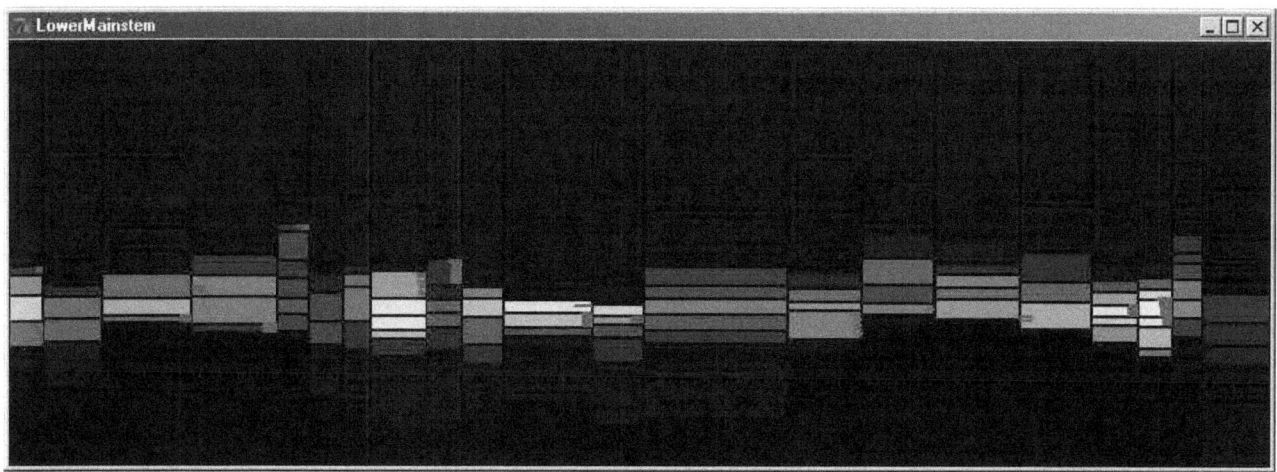

Figure 55—Animation window. One such window is opened for each habitat reach, with the reach name used as the window's title (top left).

effective.) In each cell, the lines representing fish appear with their "head" at the upstream (right) cell boundary, sorted top to bottom in descending length order.

Redds appear as an oval on the left side of their cell. Redds are also color-coded by species. Figure 55 shows two redds, one in transect 4 (the fourth vertical row of cells from the left) and one in transect 17.

22.5.1. Probe displays—

The habitat cells, fish, and redds can be "probed" from the animation window, allowing them to be examined in detail and even altered. To open a probe display to a cell (fig. 56), click on it with the left mouse button. A right mouse button click on a cell opens probe displays for all the fish and redds in the cell (fig. 57). It is usually desirable to pause execution with the model run controller's "stopActivity" button before attempting to open probe displays, but once open they can stay open when execution is resumed, so the user can see how variables change over time. Probe displays should be closed by clicking on the red button near its upper right corner.

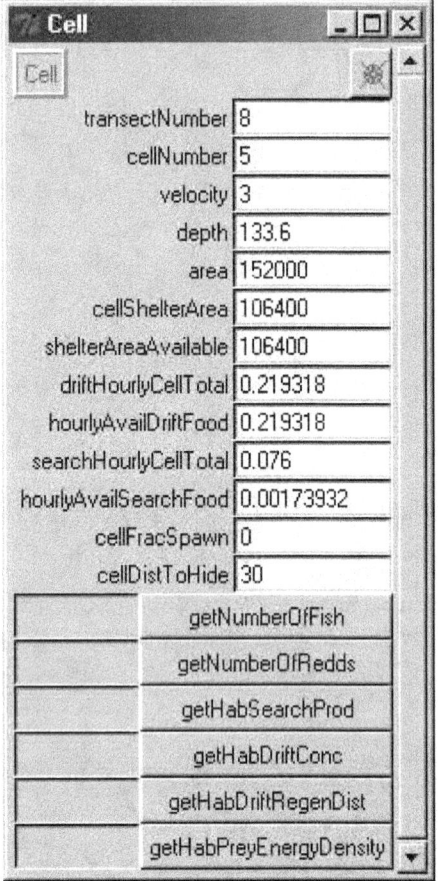

Figure 56—Cell probe display.

Probe displays include two kinds of probes: variable probes allow the user to see and change the value of a particular variable, and method probes allow the user to execute one of the object's methods (a piece of its program that executes some particular function, similar to a subroutine).

Variable probes are in the upper part of the display (*transectNumber* to *cellDistToHide* in figure 56); variable values show up in little windows to the right of the variable name. The value for a variable can be replaced manually by entering it into the value window over the old value and then hitting "Enter" (but see the note below about when values are updated).

Method probes are in the lower part of the display probes ("getNumberOfFish" to "getHabPreyEnergyDensity" in figure 56) and have an (initially) blank value window to the left of the method name. A method can be executed by clicking on the button displaying the method's name. When clicked on, these method probes return a value in the window to the left of the button.

Some method probes execute methods that require a parameter (an input to the method). These probes appear as a button with a window on its right (where the parameter value is entered before clicking the button) and with a window on its left, where the value returned from the method is displayed. The stepUntil button in the Model Run Controller (fig. 52) is an example: the required parameter is the time step number at which the model should stop again, which must be entered in the box to the right of the "stepUntil" button.

Note that the animation window is updated **after** habitat and fish simulations are completed each time step, and **before** habitat variables (depth, velocity, temperature, flow) are updated for the next time step. Therefore, any time-variable cell variables changed via probes will be reset and overwritten before the next fish simulations. For example, using the probe to change a cell's depth or velocity will have no effect because these variables are updated with a new daily value as soon as execution is resumed.

The default probe display for a trout (fig. 57A) includes four method probes that do the following things:

- "tagFish" turns the probed trout a different color in the animation window, so it can be followed when simulations are resumed. (This color is set in the Model.Setup file.)
- "tagCellsICouldMoveTo" temporarily highlights the cells that are potential movement destinations of the trout (as defined in chapter 2). The trout's current cell is not highlighted.
- "makeMeImmortal" causes the probed trout to be exempt from mortality for the rest of the simulation. None of the trout's behavior is changed.
- "killFish" causes the trout to die immediately, via the "demonic intrustion" mortality source, and "killFish" trumps "makeMeImmortal."

A

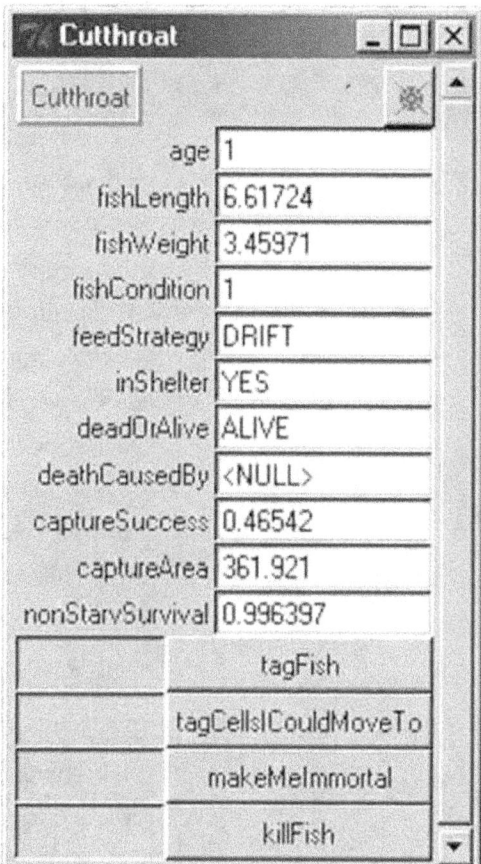

B

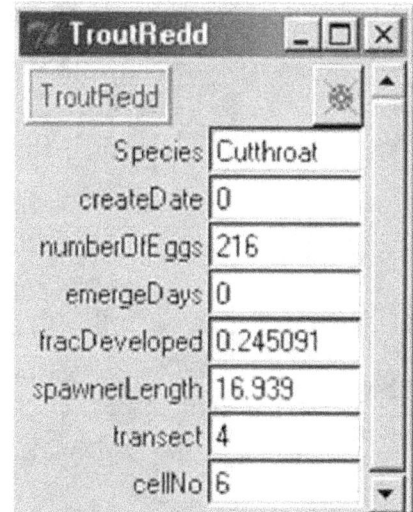

Figure 57—Trout (A) and redd (B) probe displays.

The probe displays that can be opened from the animation window include only a few selected variables and methods. (Variables and methods can be added to or removed from these displays by editing the code in the method "buildProbesIn" or—for trout—"buildFishProbes," in the code file "TroutObserverSwarm.m".) However, a "complete probe display" that shows all of an object's variables and methods can be opened by right-clicking on the box, at the top left of a probe display window, that states the object's class (either "Cell," "TroutRedd," or the trout species name). These complete probe displays, like all others, are closed by clicking on the red button at their top right corner.

For a trout, this complete probe display is empty (fig. 58A) because the methods and variables for a trout are in the Trout superclass, not the subclass for each species. However, all the complete probe display windows include a button, at the top right corner of the window, showing two green boxes and an arrow from the lower box to the upper (fig. 58A). This button opens a complete probe display of the object's superclass. Therefore, to open a complete probe display for a trout (fig. 58B), right-click on the blue species name box; then, in the newly opened window, click on the green superclass button.

22.5.2. Movies of the animation window—
It is relatively easy to make a digital "movie" of the animation window (or full screen) during a model run; these movies can be used on Web sites, in presentations, or in digital appendices to publications (e.g., Railsback and Harvey 2002). Movies are made by having InSTREAM write one graphics output file for each time step, then assembling these files into an animation file.

By default, InSTREAM produces the graphics output by capturing the animation window. If multiple habitat reaches are being simulated, the animation window for the first reach defined in the Reach.Setup file (section 20.2.3) is captured. Alternatively, the entire screen can be captured, as described below. The steps are:
- In the observer setup file, set the parameter *takeRasterPictures* to YES. This will cause InSTREAM to write, at the end of each daily time step, a new graphics file that contains an image of the animation window. The files are in Portable Network Graphics (.png) format, and have sequential file names ("Model001_Frame001.png," "Model001_Frame002.png," etc.).
- Run the model in graphics mode. To protect the animation output, the animation window now remains on top of any other windows.
- Usually, the model should be stopped after 50 to 100 time steps to produce a reasonable number of frames for a movie.

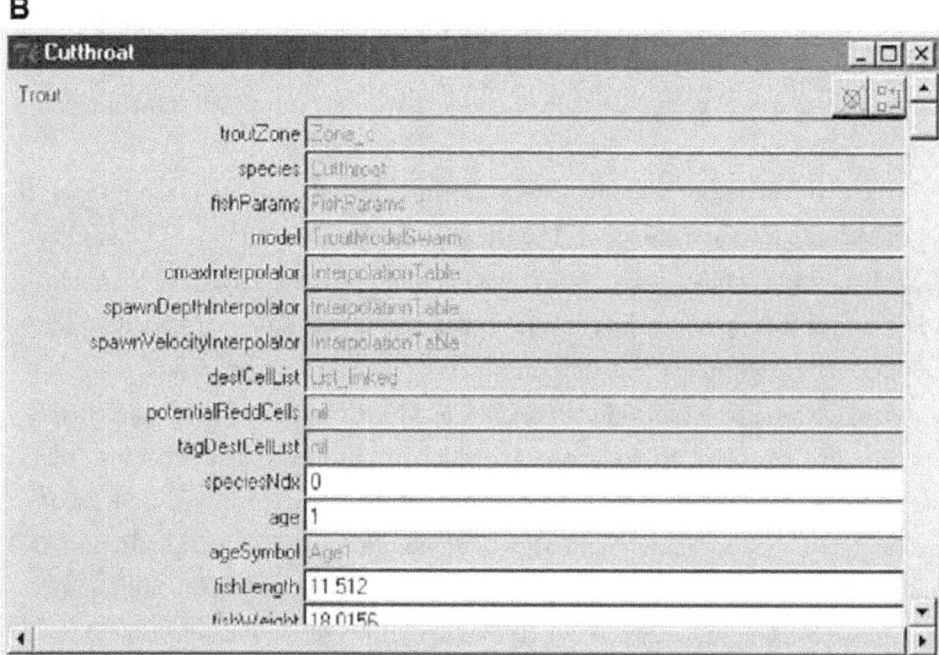

Figure 58—Right-clicking on the species name box in a trout probe display opens an empty complete probe display (A). Clicking on the green button opens a complete probe display for the trout (B).

- Animation software is used to assemble the frames into a movie. GIF Construction Set Professional, for example, is inexpensive shareware that allows the .png files to rapidly be assembled into movies in either .gif of .avi format.

A simple code change causes InSTREAM to write out pictures of the entire screen, not just the animation window. In the method writeFrame of source file TroutObserverSwarm.m, simply comment out the statement "[pixID setWidget: raster];".

23. File Output

This section describes the main output files that are written for all model runs. These output files provide the kind of detailed summary statistics that are usually most useful for understanding what happened during a simulation.

23.1. General Information on Output Files

This section briefly describes characteristics common to the main output files. All output files are written in ASCII ("plain text"). (See section 20.1.1 concerning ASCII files.)

Statistical and graphical analysis of output files is almost always required. Hence, the files were designed so they can easily be imported into spreadsheet, database, matrix math, and statistical software. Software tools such as Microsoft Access templates have been developed for analyzing some InSTREAM output files and may be available on the project Web site (http://www.humboldt.edu/~ecomodel).

The two fish output files are generated using the EcoSwarm BreakoutReporter, a generic output file-writing tool. The BreakoutReporter makes it easy to modify the software to get more, less, or different detail in the output files. By changing a few simple statements in the code, users can (1) output additional variables, (2) output different summary statistics (minimum, mean, maximum, count, sum, standard deviation, variance) for a variable, and (3) change the fish characteristics (e.g., species, age class, habitat reach) by which results are broken out or the order in which results are broken out (e.g., species first, then age versus age first, then species). The code for these output files is in the method createBreakoutReporters in the class TroutModelSwarm. (For example, code to also output the minimum and maximum length and weight of fish is currently written, but commented out, in createBreakoutReporters.) Documentation for the BreakoutReporter is at http://www.humboldt.edu/~ecomodel/software.htm.

Output files report the scenario and replicate numbers generated by the Experiment Manager (section 25). These numbers are needed to separate results from multiple model runs generated by the Experiment Manager.

If a model run stops before it is finished (because an error occurs or because it is killed by the user), the output files will persist and provide results up to when the simulation died.

Users should pay attention to the *appendFiles* parameter and how it is set in Model.Setup and Experiment.Setup. This parameter controls whether output files are overwritten vs. appended each time a model run starts (except the separate model runs within an experiment generated by the Experiment Manager).

The parameter *fileOutputFrequency* in Model.Setup controls how often output is written to the two fish output files. Using a higher value of *fileOutputFrequency* can reduce execution time (because generating the fish outputs using BreakoutReport requires quite a few computations) and reduce the volume of output to be stored and analyzed.

23.2. Output File Descriptions

The main output files are described here. New users can most easily view these files by opening them in spreadsheet software.

Live fish output file. This file reports a time series of summary statistics on the live trout population. The user provides the file name, via the parameter *fishOutputFile* in the model setup file (section 20.2.4). By default, this file provides the abundance, mean weight, and mean length of the trout population, broken out by habitat reach, trout species, and age class.

Fish mortality output file. This file reports mortality statistics: the number of trout that have died via each mortality source. Results are broken out by habitat reach, species, and age class. On each output date, a line is written to this file reporting the **cumulative** number of trout that have died of each type of mortality: the total deaths since the start of the simulation. These results can be used to produce graphs (e.g., fig. 59) useful for understanding when different causes of mortality are important.

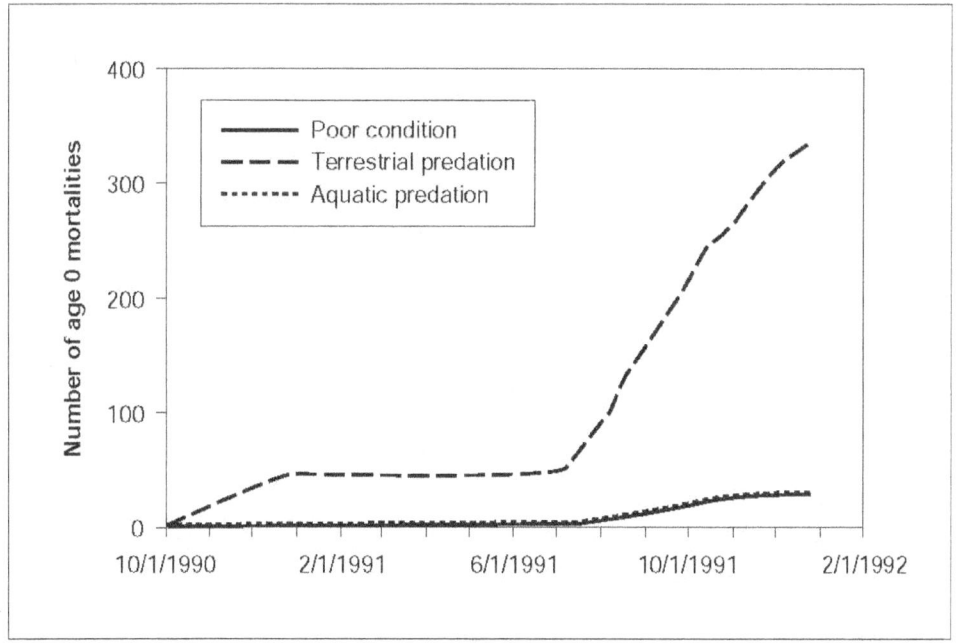

Figure 59—Example plot of results from the fish mortality output file, showing the cumulative mortality of age 0 trout from the three most important sources. (There is no mortality between January 1 and about July 1 because there are no age 0 trout during that period.)

Redd output file. This file provides one line of results for each redd, written on the day when the redd becomes empty (all eggs have died or turned into new trout). The user specifies the file name, via the parameter *reddOutputFile* in the model setup file. Results reported for each redd include the spawner's length, weight, and age; the dates the redd was created and emptied; the redd's location (reach, transect, and cell); the initial number of eggs; and the number of eggs dying of each redd mortality source and the number emerging as new trout. The output also includes a "ReddID," which is simply a unique alphanumeric code for each redd.

Redd mortality output file. A daily egg mortality report is provided for each redd. The name of this file is also specified in Model.Setup. When each redd is empty, it writes its report to the end of this file. The report includes the ReddID code, the initial number of eggs, and (on separate lines) the number of eggs dying of each mortality source on each day of the redd's existence.

Habitat output files. These files output the values of the time-series habitat variables. One file is written for each habitat reach; its name is the reach's name with "Habitat.out" appended to it (e.g., MainstemReachHabitat.out). Each line reports the date and the daily flow, temperature, and turbidity values (which were input via the files described in section 20.4.4).

These habitat output files are actually optional (section 24) and can be suppressed by commenting out the statement `#define HABITAT _ REPORT _ ON` in the code file HabitatManager.h.

Depth use histogram output. This file is simply a file version of the depth histogram described in section 22.4. Its file name is always FishDepthHisto.out. The file (generated by Swarm's histogram tool) contains no column headings, and includes one line for each simulation day. Each line includes values, separated by tab characters, equal to the number of fish in each histogram bin. Note that outliers (trout using depths greater than the histogram's upper bound) are ignored in this file, so the total number of trout on a line of output may be less than the total number of trout alive. (More useful output for analysis of habitat availability and habitat use by model trout is provided by the optional CellFishInfo.rpt output file described in section 24.)

This file is not generated when InSTREAM is executed in batch mode (section 19.4) and is always overwritten each model run (including each replicate of each scenario if the Experiment Manager is used).

Velocity use histogram output. This is file output of the velocity histogram described in section 22.4, and always named FishVelocityHisto.out. The file format and characteristics are the same as those of the depth histogram.

24. Optional Output Files for Testing and Specialized Studies

The InSTREAM software includes a number of optional output files that are normally not written, but can be turned on when more detailed output is desired. These "reports" are often useful for testing changes in the software or parameters, and for supporting more detailed analysis of trout behavior. However, many of these optional output files can be extremely large and writing them can make InSTREAM much slower to execute.

The optional output files are turned on by making very simple edits in several of the code files. Normally, the statement that activates each of these files is commented out by placing two slashes in front of it. These statements appear in several interface (.h) source code files, and include the word #define and the report name. For example, a report providing details on redd survival can be turned on by editing the file "TroutModelSwarm.h" and changing this line:

```
//#define REDD_SURV_REPORT
```

to:

```
#define REDD_SURV_REPORT
```

Then the source code must be recompiled. When changes are made to any interface files, it is best to recompile using make clean first, then make (section 19.3).

Table 32 provides a complete list of the optional output files. The first column tells where in the source code—which statement in which .h file—the file is activated. The second column provides the name of the output file. Some of the optional reports produce a separate output file for each habitat reach, in which case the file name starts with the reach name. The third column describes the information provided in the file.

With one exception, these optional output files are not controlled by the Model. Setup parameter *appendFiles*; instead, they are always overwritten at the start of each model run. Even for multiple model runs generated by the Experiment Manager (section 25), these files will report results only from the last model run.

The exception is the CellFishInfo.rpt output (the last line of table 32). This output is controlled by *appendFiles*, and is also designed so that results from multiple runs generated by the Experiment Manager are automatically appended.

Table 32—Optional output files

Report file name	Source code file and activating statement	Description of report
FoodAvailability.rpt	Cell.h #define FOODAVAILREPORT	Produces one line each time a fish moves into a cell. Reports the cell's hourly food production rate, the hourly rate of food availability (unused by fish already in the cell), and the amount consumed by the fish. Results are separate for drift and search food types.
(reachname)Cell_Flow_-Depth_Test.rpt	HabitatManager.h #define DEPTH_REPORT_ON	Produces one file for each habitat reach. One line is written at the start of each day, reporting the flow, and depth in each cell.
(reachname)Cell_Flow_-Velocity_Test.rpt	HabitatManager.h #define VELOCITY_REPORT_ON	Like the depth report, but cell velocity is reported.
(reachname)CellDepth-AreaVelocity.rpt	HabitatManager.h #define DEPTH_VEL_RPT	Produces one file for each reach. On each day, one line is written for each habitat cell with depth above zero; cell area, depth, and velocity are output.
MoveTest.rpt	Trout.h #define MOVE_REPORT_ON	Writes one line when each trout selects the habitat cell it will occupy, each day. Reports the data (about the selected cell and the trout) used to select this best cell. Also reports intermediate calculations in the trout's movement decision (e.g., its respiration rate, net energy intake). Note: The output is first produced when the fish are initialized on the first simulated day, when fish are not yet sorted in size order.
Ready_To_Spawn.rpt	Trout.h #define READY_TO_SPAWN_RPT	Produces one line per day for each female trout. Reports whether the female determined it was ready to spawn and the variables used to make the determination.
Spawn_Cell.rpt	Trout.h #define SPAWN_CELL_RPT	Produces one line of output each time a female trout spawns and builds a redd. For each potential spawning cell, reports the cell depth, velocity, and fraction of spawning gravel; and the calculated variables used to rate spawning cells: depth and velocity suitability, overall spawning quality.
ReddSurvivalTest.rpt	TroutModelSwarm.h #define REDD_SURV_REPORT	Produces a report on egg survival for a redd on the day when the redd has no more live eggs (owing to mortality and emergence). Reports the redd's species, location (transect and cell numbers), and initial number of eggs. Then, for each day of the redd's existence, a line reports the temperature, flow, depth, and number of eggs dying of each mortality ource.
(reachname) CellFishInfo.rpt	TroutModelSwarm.h #define PRINT_CELL_FISH_REPORT	Produces one file for each reach. On each day, writes one line for each habitat cell. Reports cell area, depth, velocity, distance to hiding cover, fraction with velocity shelter, and statistics on the fish in the cell.[a]

[a] This file is produced using the EcoSwarm BreakoutReporter, and is very similar to the live fish output file. The fish statistics provided in this file normally include the number of fish in the cell, broken out by species and age class. However, these statistics can be modified (section 23.1), by altering the BreakoutReporter code in method buildCellFishInfoReporter in file HabitatSpace.m

25. Experiment Manager

All users of InSTREAM need to at least be aware of the Experiment Manager because it has powerful features that are often extremely useful, but it can also act like a nightmarish bug if ignored. These features overwrite values provided in setup and parameter files, causing the model's behavior to be inexplicable if one forgets to pay attention to the Experiment Manager. But any serious user of InSTREAM will quickly learn to depend on the Experiment Manager to conduct simulation experiments quickly, easily, and reliably.

25.1. What the Experiment Manager Does

The purpose of the Experiment Manager is to allow users to set up and execute simulation experiments that use multiple model runs with different inputs, without having to modify either the software or the parameter files. With a few simple statements in its setup file, a complex experiment can be set up to run automatically. The experiments can include both **scenarios** and **replicates**. A scenario is a single set of inputs and parameter values; different scenarios are defined by specifying which inputs or parameters differ among them. Replicates are repeated runs of the same scenario, with only the pseudorandom numbers (which primarily affect mortality; see chapter 2) differing among replicates.

The Experiment Manager has two functions. The first is to set up and execute the number of simulations specified in its setup file. For example, the user may set up the Experiment Manager to define 10 scenarios (perhaps five different values for some parameter, for each of two flow input files) and three replicates. The Experiment Manager would determine that 30 model runs are needed and start one run after another. The second function is to modify the parameters for each of the model runs to implement the scenarios. This function occurs at the start of a model run: the Experiment Manager stops the model after parameter values have been read in, then overwrites the value of any parameters that are specified in its setup file. For replicates, the Experiment Manager simply changes the random number seed (the seed value provided in Model.Setup is multiplied by the replicate number). The Experiment Manager then re-starts the model and has no more effect until the next run starts. (To be precise: the TroutModelSwarm method instantiateObjects is executed; then the Experiment Manager is executed and modifies parameter values; then the TroutModelSwarm method buildObjects is executed and the simulation proceeds.)

The Experiment Manager's scenarios almost always involve modifying the value of variables that are in setup or parameter files. The Experiment Manager is generally **not** useful for directly modifying variables that are in input files (flows, temperatures, cell characteristics, etc.); however, it works very nicely to define such scenarios by creating different input files and using the Experiment Manager to control which input file is used.

The following sections describe the setup file that controls the Experiment Manager, and provide many examples that can be followed to set up experiments.

25.2. General Procedure for Setting Up Experiments

The general procedure for using the Experiment Manager is to (1) specify the scenarios and replicates in the Experiment.Setup file, (2) run InSTREAM in batch (non-graphics) mode, and (3) examine and analyze the results. Results are usually written to one file, with output labeled by scenario and replicate number. It is always good to archive a copy of the Experiment.Setup file with the output files from an experiment, to document exactly what scenarios were executed.

25.2.1. Experiment.Setup format—

The Experiment.Setup file controls the Experiment Manager. This file is always read by InSTREAM, so it must be configured correctly even if the model is to be run with no automated experiments. This format of this setup file is illustrated in the following example and described below (see additional examples in section 25.3).

```
Experiment setup file -
Created Feb 21 2005
For demo example

numberOfScenarios              3
numberOfReplicates             5

sendScenarioCountToParam:      scenario
inClass:                       TroutModelSwarm

sendReplicateCountToParam:     replicate
inClass:                       TroutModelSwarm

ClassName                      FishParams
InstanceName                   Cutthroat
ParamName                      fishFitnessHorizon
ValueType                      double
Value                          60.0
Value                          90.0
Value                          120.0
```

Experiment.Setup contains:

- Comments, which can be included anywhere as lines that start with the character "#".
- Three header lines, not used by the computer. These can be up to 200 characters long.
- A blank line
- Two lines on which (a) the number of scenarios and (b) the number of replicates for each scenario are specified. These values must be at least 1, and are not limited by the software.
- A blank line
- Two lines that provide the variable name and class to which the code sends the current scenario count, during model execution. These lines should not be changed.
- A blank line
- Two lines that provide the variable name and class to which the code sends the current replicate count. These lines should also not be changed.
- A blank line

Following these initial blocks of text, the file contains zero or more additional blocks, which each specify a model parameter to be varied among scenarios and the value it has for each scenario. (At least one such block must be specified if the number of scenarios is greater than one.) There is no limit to how many of these blocks can be specified, or how many parameters can be controlled by the Experiment Manager. These blocks contain the following lines; blocks are separated by a blank line.

- The word ClassName followed by the name of the class in which the parameter value is defined. "Class" refers to the object-oriented software structure in which each object in the model is an instance of a particular class. (Class, instance, and variable names are further explained in section 25.2.2.)
- The word InstanceName followed by the name of the instance of the class for which the parameter is to be varied. If the word NONE is provided for InstanceName, then the parameter is varied for all instances of the class.
- The word ParamName followed by the name of the parameter to be varied.
- The word ValueType followed by the kind of value that the parameter contains. The value type must be one of the types defined in section 25.2.3.
- The word Value followed by the parameter's value for the first scenario. This line is repeated for each scenario: there must be one value provided for each scenario, even if the value is the same for some scenarios.

25.2.2. Class and instance names for typical experiments—

The Experiment Manager is highly flexible, allowing users access to any instance variable of any class in InSTREAM. However, using it successfully for unusual experiments requires detailed knowledge of the software; some inputs cannot be usefully manipulated because they have already been used before the Experiment Manager executes (e.g., the hydraulics data files) or because their values are over-written after the Experiment Manager executes (e.g., reach flow, cell depth and velocity). Table 33 describes the usual applications of the Experiment Manager that can be made with confidence.

Input that **cannot** be manipulated by the Experiment Manager includes parameters in the Observer.Setup and Species.Setup files and the hydraulics data file names in Reach.Setup.

Table 33—Class and instance names for parameters commonly used in the Experiment Manager

Parameters	ClassName	InstanceName
Model setup parameters (any parameters in Model.Setup)	TroutModelSwarm	NONE
File names for cell data, flow, temperature, and turbidity input (parameters flowFile, temperatureFile, turbidityFile in the reach setup file)	HabitatSpace	The reach name specified in Reach.Setup (or NONE if only one reach is simulated)
Habitat parameters (any parameter in a habitat parameter file)	HabitatSpace	The reach name specified in Reach.Setup (or NONE if the Experiment Manager is to alter the parameter for all habitat reaches)
Trout and redd parameters (any parameter in a trout parameter file)	FishParams[a]	The species name specified in Species.Setup (or NONE if the Experiment Manager is to alter the parameter for all species)

[a] Trout parameter values are not stored in the model trout themselves, but in a separate "FishParams" object for each species.

25.2.3. Valid parameter value types—

The Experiment.Setup file must provide a "ValueType" for each parameter to be manipulated. Valid value types are defined in table 34. The value type depends on the parameter itself. If users are not sure what type a parameter is, they should find the parameter in the header (.h) file for the parameter's class. For example, the parameter controlled by the example setup file in section 25.2.1 is *fishFitnessHorizon*, the number of days over which a fish evaluates its habitat selection decision. This parameter could be either an integer or floating point (real) number. Searching the file FishParams.h for *fishFitnessHorizon* finds the declaration "const double *fishFitnessHorizon*," indicating that this parameter is a double-precision floating point number, so ValueType should be "double."

Table 34—Value types for the Experiment.Setup file

Value type[a]	Definition
BOOL	A boolean variable, with a value of either YES or NO (all upper case)[b]
date	A date variable in MM/DD/YYYY format
day	A day-of-the-year variable in MM/DD format
double	A double-precision floating point variable (any number that is not an integer)
filename	The name of an input file (a character string)
int	An integer variable

[a] The "ValueType" field in the Experiment.Setup file must **exactly** match one of the values in this column, including upper/lower case.
[b] Note that boolean variables must have values of 0 or 1 in other setup files, but values of either YES or NO in Experiment.Setup.

25.2.4. Using instance names—

The InstanceName field in Experiment.Setup allows the user to manipulate parameters for one instance of a class: for one of several trout species, or for one of several habitat reaches (see table 33). InstanceName is set to NONE if there is only one instance per class, or if the same parameter change is to be made for all instances.

Separate parameter blocks can be used in the Experiment.Setup file to change the same parameter different ways for different species or habitat reaches. The following examples illustrate how the Experiment Manager uses instance names. The examples assume a version of InSTREAM with three trout species named Cutthroat, Rainbow, and Brown; and that only one scenario is generated. Example 1 changes the value of parameter *fishEnergyDensity* to 4555, but only for cutthroat trout; other species retain the value in their parameter files:

```
ClassName       FishParams
InstanceName    Cutthroat
ParamName       fishEnergyDensity
ValueType       double
Value           4555
```

Example 2 changes the value of *fishEnergyDensity* for all species:

```
ClassName       FishParams
InstanceName    NONE
ParamName       fishEnergyDensity
ValueType       double
Value           4555
```

219

Example 3 changes the value of *fishEnergyDensity* to 4555 for cutthroat trout and to 5000 for brown trout; rainbow trout are unaffected:

```
ClassName       FishParams
InstanceName    Cutthroat
ParamName       fishEnergyDensity
ValueType       double
Value           4555

ClassName       FishParams
InstanceName    Brown
ParamName       fishEnergyDensity
ValueType       double
Value           5000
```

Example 4 changes *fishEnergyDensity* to 4555 for cutthroat trout and to 5000 for all other species. The order in which these two blocks appear does not matter—changing parameter values for a specific instance always overrides a general change (via `InstanceName NONE`) to all instances.

```
ClassName       FishParams
InstanceName    Cutthroat
ParamName       fishEnergyDensity
ValueType       double
Value           4555

ClassName       FishParams
InstanceName    NONE
ParamName       fishEnergyDensity
ValueType       double
Value           5000
```

Example 5 changes *fishEnergyDensity* to 5000. It does not cause an error even though *fishEnergyDensity* is included twice—if the same parameter is included several times in Experiment.Setup in the same way, the last value is used.

```
ClassName       FishParams
InstanceName    Cutthroat
ParamName       fishEnergyDensity
ValueType       double
Value           4555

ClassName       FishParams
InstanceName    Cutthroat
ParamName       fishEnergyDensity
ValueType       double
Value           5000
```

25.2.5. Controlling where output goes—

Normally the Experiment Manager is set up so output from all scenarios and replicates are sent to the same output files, which are automatically appended for each model run (even if the parameter *appendFiles* is set to 0 in Model.Setup). The standard output files include the scenario and replicate number for all output.

An alternative is to include unique names for output files for each scenario in the Experiment.Setup file. This would write results from each scenario to a different output file (but multiple replicates would still be in the same file). (This capability has not been tested and no examples are provided.)

25.2.6. Checking the Experiment Manager—

The Experiment Manager includes several kinds of error checking; any of the following stop execution and generate an error statement:

- The number of scenarios or replicates is set to zero.
- A ValueType field has an invalid value, or its value does not match that of the parameter.
- A ClassName or InstanceName field has an invalid value.
- The named parameter does not exist in the specified class.
- The number of values provided for a parameter differs from the number of scenarios.

Several procedures allow verification that the Experiment Manager produced the intended parameter values. Manipulations of input data (e.g., the flow or temperature input file) can be checked by examining the habitat output files (section 23.2).

Habitat parameters can be checked by running InSTREAM in graphics mode and using the HabitatSpace probe display (section 22.5.1). (Remember that multiple scenarios can be run in graphics mode by clicking on "Start" on the **main** control panel after each run finishes and that **all** habitat variables can be viewed by right-clicking on the box labeled HabitatSpace in the top left corner of the probe display; figure 53.)

Trout parameters can be checked by turning on an optional output that prints out trout parameter values at the start of each model run (after they have been manipulated by the Experiment Manager). A separate output file is created for each species; these are named SpeciesXParamCheck.out, where X is the name of the trout species. There are two ways to turn this optional output on; one is to add this line to Model.Setup:

```
printFishParams 1
```

The second way is by controlling this parameter via the Experiment.Setup file, using a block such as this:

```
ClassName        TroutModelSwarm
InstanceName     NONE
ParamName        printFishParams
ValueType        BOOL
Value            NO
Value            YES
```

This example prints out fish parameters only at the start of the second scenario. **Note that this output is overwritten each model run, so it only reflects the last scenario started.**

25.3. Example Experiment.Setup Files

The Experiment Manager is fairly complicated, so a number of examples are provided here. Most users should be able to design the experiments they need by modifying these Experiment.Setup files.

25.3.1. Single model runs—

When users want to run a single model run, with no parameters altered by the Experiment Manager, the following Experiment.Setup file can be used.

```
Experiment setup file -
Created Feb 21 2005
Example for deactivated Experiment Manager

numberOfScenarios            1
numberOfReplicates           1

sendScenarioCountToParam:    scenario
inClass:                     TroutModelSwarm

sendReplicateCountToParam:   replicate
inClass:                     TroutModelSwarm
```

25.3.2. Replicate simulations—

Users often want to replicate a single scenario to understand the stochasticity of model results. This Experiment.Setup will run the selected number of replicates (five, in this example) and append results from each model run to the output files.

```
Experiment setup file -
Created Feb 21 2005
Example for replication of one scenario

numberOfScenarios          1
numberOfReplicates         5

sendScenarioCountToParam:  scenario
inClass:                   TroutModelSwarm

sendReplicateCountToParam: replicate
inClass:                   TroutModelSwarm
```

25.3.3. Parameter sweep for trout—

One of the most common kinds of experiment is a "parameter sweep": an experiment in which one parameter is varied over a wide range. The example uses the trout parameter *mortFishTerrPredMin*, and applies the sweep to all trout species in the model. Therefore, this example could be used to show how trout populations respond to increasing levels of terrestrial predation risk (decreasing survival probability).

```
Experiment setup file -
Created Feb 21 2005
Example trout parameter sweep

numberOfScenarios          6
numberOfReplicates         1

sendScenarioCountToParam:  scenario
inClass:                   TroutModelSwarm

sendReplicateCountToParam: replicate
inClass:                   TroutModelSwarm

ClassName                  FishParams
InstanceName               NONE
ParamName                  mortFishTerrPredMin
ValueType                  double
Value                      0.950
Value                      0.961
Value                      0.972
Value                      0.983
Value                      0.994
Value                      1.0
```

25.3.4. Habitat parameter sweep—

This parameter sweep varies a habitat variable: the drift food concentration in one habitat reach (MainstemUpperReach). This example illustrates an experiment to see how trout populations respond to food availability in one of several reaches. This example also illustrates that parameter values can be in scientific notation.

```
Experiment setup file -
Created Feb 21 2005
Example habitat parameter sweep

numberOfScenarios            6
numberOfReplicates           1

sendScenarioCountToParam:    scenario
inClass:                     TroutModelSwarm

sendReplicateCountToParam:   replicate
inClass:                     TroutModelSwarm

ClassName                    HabitatSpace
InstanceName                 UpperMainstem
ParamName                    habDriftConc
ValueType                    double
Value                        5.0E-11
Value                        7.0E-11
Value                        9.0E-11
Value                        1.10E-10
Value                        1.30E-10
Value                        1.50E-10
```

25.3.5. Multiple parameter sweep—

This experiment explores interactions among variables: What happens if two parameters are varied such that all combinations are simulated? This example varies two parameters (for fish vs. terrestrial predation risk), with three values of each; but the same approach can be used with more parameters and more values.

```
Experiment setup file -
Created Feb 21 2005
Example multiple parameter sweep

numberOfScenarios            9
numberOfReplicates           1

sendScenarioCountToParam:    scenario
inClass:                     TroutModelSwarm
```

```
sendReplicateCountToParam:    replicate
inClass:                      TroutModelSwarm

ClassName                     FishParams
InstanceName                  NONE
ParamName                     mortFishAqPredMin
ValueType                     double
Value                         0.95
Value                         0.95
Value                         0.95
Value                         0.97
Value                         0.97
Value                         0.97
Value                         0.99
Value                         0.99
Value                         0.99

ClassName                     FishParams
InstanceName                  NONE
ParamName                     mortFishTerrPredMin
ValueType                     double
Value                         0.95
Value                         0.97
Value                         0.99
Value                         0.95
Value                         0.97
Value                         0.99
Value                         0.95
Value                         0.97
Value                         0.99
```

25.3.6. Alternative daily input files—

A primary application of InSTREAM is to compare alternative instream flow and temperature scenarios. This example illustrates how three streamflow regimes can be contrasted by creating three alternative flow input files and using the Experiment Manager to generate replicate simulations of each. (The flow input files could, for example, be generated by a reservoir model simulating alternative reservoir operating rules.) This example applies the same flow input to all habitat reaches, including when only one reach is simulated.

Note that the Reach.Setup file (section 20.2.3) contains the input files names for flow, temperature, and turbidity (flowFile, temperatureFile, turbidityFile), and these are passed to the habitat reach objects (class HabitatSpace) before the Experiment

Manager is activated. Therefore, the Experiment Manager accesses these file names in HabitatSpace.

```
Experiment setup file - Comparison of alternative flow scenarios
Created Feb 21 2005
Three flow scenarios, five replicates of each

numberOfScenarios              3
numberOfReplicates             5

sendScenarioCountToParam:      scenario
inClass:                       TroutModelSwarm

sendReplicateCountToParam:     replicate
inClass:                       TroutModelSwarm

ClassName                      HabitatSpace
InstanceName                   NONE
ParamName                      flowFile
ValueType                      filename
Value                          FlowScenario1.Data
Value                          FlowScenario2.Data
Value                          FlowScenario3.Data
```

25.3.7. Alternative initial populations—

This example shows how to initialize the model with different trout populations. Varying the initial population characteristics could be useful, for example, to determine how long effects of initial conditions persist (e.g., after three simulation years, are results independent of the initial abundance and size distribution?).

The input that specifies initial population characteristics is in the initial population data file (section 20.4.3) and cannot be directly manipulated by the Experiment Manager. However, the Model.Setup file includes the parameter *popInitDate*, which specifies which data in the initial population data file are used to initialize a model run. This experiment is conducted by editing the initial population data file so it includes two complete, alternative sets of initial conditions, each with their own initialization date, as illustrated in section 20.4.3. Then the Experiment Manager is used to choose the initialization dates to actually start the model runs. This example includes five replicates of the two initial population scenarios.

This example is also easily modified to manipulate other parameters in Model.Setup.

```
Experiment setup file - for alternative initial populations
Created Feb 21 2005
Example for changing model setup parameters

numberOfScenarios              2
numberOfReplicates             5

sendScenarioCountToParam:  scenario
inClass:                   TroutModelSwarm

sendReplicateCountToParam: replicate
inClass:                   TroutModelSwarm

ClassName                  TroutModelSwarm
InstanceName               NONE
ParamName                  popInitDate
ValueType                  date
Value                      10/1/1999
Value                      10/1/2000
```

25.3.8. Alternative cell data files—

Multiple cell data files could be used to simulate different availabilities of hiding or feeding cover or spawning gravel. The cell data file is a variable of the Habitat-Space class, and the instance name is the name of the habitat reach as defined in Reach.Setup.

```
Experiment setup file - for alternative cell data files
Created Apr 20, 2005
Example

numberOfScenarios              2
numberOfReplicates             5

sendScenarioCountToParam:  scenario
inClass:                   TroutModelSwarm

sendReplicateCountToParam: replicate
inClass:                   TroutModelSwarm

ClassName                  HabitatSpace
InstanceName               MiddleReachBearCreek
ParamName                  cellDataFile
ValueType                  filename
Value                      HiCoverCell.Data
Value                      LoCoverCell.Data
```

25.3.9. Year shuffler scenarios—

The year shuffler facility in InSTREAM (described in chapter 2) can generate scenarios that differ only by having years of input data shuffled. A separate random number generator is used to shuffle years, allowing year randomization to be controlled separately from other stochastic processes. The following example shows how to generate five scenarios that differ only in the sequence in which years of input occur.

```
Experiment setup file -
For year shuffling scenarios
SFR 5/3/05

numberOfScenarios              5
numberOfReplicates             1

sendScenarioCountToParam:      scenario
inClass:                       TroutModelSwarm

sendReplicateCountToParam:     replicate
inClass:                       TroutModelSwarm

ClassName                      TroutModelSwarm
InstanceName                   NONE
ParamName                      shuffleYearSeed
ValueType                      int
Value                          7377
Value                          7378
Value                          7379
Value                          7370
Value                          7371
```

25.3.10. Year shuffling as replication—

The year shuffler can be used as an alternative way to replicate scenarios: a scenario can be run multiple times with different random reordering of the input data to see how robust its results are to the sequence of input years. This example shows how to generate five such year-shuffler replicates of three scenarios. The three scenarios are alternative daily flow files that each represent a minimum flow requirement at a diversion dam.

The year-shuffler replicates must be treated by the Experiment Manager as separate scenarios because they change the year-shuffler random number seed, not the seed used for all other stochastic processes.

```
Experiment setup file -
For instream flow experiments
SFR 5/3/05

numberOfScenarios          15
numberOfReplicates         1

sendScenarioCountToParam:  scenario
inClass:                   TroutModelSwarm

sendReplicateCountToParam: replicate
inClass:                   TroutModelSwarm

ClassName                  HabitatSpace
InstanceName               NONE
ParamName                  flowFile
ValueType                  filename
Value                      Divers01Flow.Data
Value                      Divers01Flow.Data
Value                      Divers01Flow.Data
Value                      Divers01Flow.Data
Value                      Divers01Flow.Data
Value                      Divers02Flow.Data
Value                      Divers02Flow.Data
Value                      Divers02Flow.Data
Value                      Divers02Flow.Data
Value                      Divers02Flow.Data
Value                      Divers03Flow.Data
Value                      Divers03Flow.Data
Value                      Divers03Flow.Data
Value                      Divers03Flow.Data
Value                      Divers03Flow.Data

ClassName                  TroutModelSwarm
InstanceName               NONE
ParamName                  shuffleYearSeed
ValueType                  int
Value                      7377
Value                      7378
Value                      7379
Value                      7370
Value                      7371
```

```
Value                            7377
Value                            7378
Value                            7379
Value                            7370
Value                            7371
Value                            7377
Value                            7378
Value                            7379
Value                            7370
Value                            7371
```

26. Troubleshooting Guide

This section lists symptoms and solutions for common problems encountered installing and running InSTREAM. If you have mysterious problems that are not described here, you should still make sure you have made none of the mistakes described in this guide: sometimes small mistakes (e.g., a blank within a text variable such as a file name or reach name) have strange effects. Additional help for Swarm models in general is available from the Frequently Asked Questions resources on the Swarm Web site, http://www.swarm.org.

Symptom: When I try to compile the model, Cygwin says something like "Makefile: ... Makefile.appl: No such file or directory. make: *** No rule to make target ..."

Solution: The environment variable "SWARMHOME" is not set correctly, most likely because you (1) forgot to set this environment variable or (2) set it to the wrong directory, perhaps because Cygwin was installed to a nonstandard directory.

If you cannot figure out how to set SWARMHOME correctly, do this instead:

(1) Open the makefile in an editor. In the directory of source code for InSTREAM will be a file named "Makefile" that contains the directions Cygwin uses to compile the model. This is a plain text (ASCII) file you can edit using Notepad or Wordpad. At the top of the makefile you should see several lines similar to:

```
ifeq ($(SWARMHOME),)
SWARMHOME=/usr
endif
```

(2) Comment those lines out by putting a "#" character in front of them, and add a new line identifying SWARMHOME's location. The result should be:

```
#ifeq ($(SWARMHOME),)
#SWARMHOME=/usr
#endif
SWARMHOME=C:/Cygwin/usr
```

where the last line points to the "usr" subdirectory under the main Cygwin directory. For example, if you installed Cygwin to "C:/Program Files" then the last line would be:

```
SWARMHOME=C:/Program Files/Cygwin/usr
```

Make sure you use forward slashes "/". See the following entry for a likely problem at this point.

Symptom: I edited the Makefile (or a code file or input file) using a Windows editor (e.g., Notepad, Wordpad, Word, Excel), and saved the change; but Cygwin and InSTREAM ignore the change.

Solution: Check whether the editor saved the changes in a new file called "Makefile.txt" instead of in the original file "Makefile." These editors sometimes insist that all plain-text files should end in ".txt". You can overcome this insistence by putting the file name in quotation marks when telling the editor where to save the file.

Symptom: When I open output files in Windows Notepad, the lines are all run together.

Solution: You installed Cygwin with the "Unix file format" option, so output files have Unix-type line end characters. Re-install Cygwin using the DOS file format option; or use the Cygwin "unix2dos" conversion facility to convert individual files to Windows format (in Cygwin, type "unix2dos filename"); or open the files using Wordpad, Word, or Excel, which can handle Unix format.

Symptom: InSTREAM says it cannot find an input file that really is there; or there is an error reported by Swarm's "ObjectLoader."

Solution: Make sure the file name is completely correct, **including upper vs. lower case.** File names (and reach names, in Reach.Setup) cannot include blanks.

If you are in Linux, was the file created in Windows? If so, you must use the "dos2unix" utility to convert it to Unix file format; otherwise, InSTREAM will not be able to read it. (But if you accidentally use dos2unix on the executable file instream.exe, it will be ruined and you will have to recompile it.)

Symptom: The model will not read a text parameter such as a file name.

Solution: See the previous line concerning file names.

Make sure there are no blanks after the parameter value.

Make sure the setup file containing the parameter is in the proper Unix or DOS file format.

Symptom: When I try to start the model by typing `instream.exe`, I only get the error "bash: instream.exe: command not found".

Solution: Type `./instream.exe` to start the model.

Symptom: I changed a parameter value in the trout (or habitat) parameter file, but the change has no effect.

Solution: Check the Experiment.Setup file to see if the parameter value is being controlled by the Experiment Manager.

Symptom: After I edit an input file in a spreadsheet, the model runs for a while then stops; the error statement says that something is wrong with dates.

Solution: See the end of section 20.4.4.

Symptom: When I start InSTREAM up, it stops while initializing a model run with an error statement saying that a date was improperly formatted. But all the dates in my input files seem to be correct.

Solution: There have been mysterious problems with InSTREAM's date/time management software on some operating systems some of the time. If using Windows, make sure all input files are in DOS file format. Contact the InSTREAM developers if the problem persists.

Symptom: When I try to start InSTREAM, I get an error "Unable to locate DLL— The dynamic link library cygwin1.dll could not be found..."

Solution: You tried to execute instream.exe from the Windows Explorer or from a Windows command prompt (DOS) window. You can only execute InSTREAM from within Cygwin.

Symptom: When I try to compile the model the first time, I get a number of error statements, including one that says "defobj/COM.h" is missing.

Solution: In older installations of Swarm, a Swarm file COM.h was missing. See InSTREAM's developers if this problem arises.

Acknowledgments

This report and the model it describes are products of collaborative research conducted by the Department of Mathematics, Humboldt State University; USDA Forest Service, Pacific Southwest Research Station, Redwood Sciences Laboratory; and Lang, Railsback and Associates. Funding was provided by the National Center for Environmental Research (NCER) STAR Program, U.S. Environmental Protection Agency, under EPA Agreement RD-83088601-0 with Humboldt State University; and by Research Joint Venture Agreements PSW-99-007-RJVA and 03-JV-11272133-025 between the Pacific Southwest Research Station and Lang, Railsback and Associates. We thank Jason Dunham, Charles Gowan, Amanda Rosenberger, and Brad Shepard for reviewing sections of this report. The InSTREAM software is developed and maintained by Steve Jackson.

This report has not been subjected to EPA review and therefore does not necessarily reflect the views of that agency, and no official endorsement should be inferred.

English Equivalents

When you know:	Multiply by:	To find:
Centimeters (cm)	0.394	Inches
Centimeters (cm)	0.0328	Feet
Centimeters per second (cm/s)	0.0328	Feet per second
Square centimeters (cm^2)	0.001076	Square feet
Cubic centimeters (cm^3)	3.53×10^{-5}	Cubic feet
Meters (m)	3.28	Feet
Meters	1.094	Yards
Square meters (m^2)	10.76	Square feet
Cubic meters (m^3)	35.3	Cubic feet
Cubic meters per second (m^3/s)	35.3	Cubic feet per second
Seconds per cubic meter (s/m^3)	0.0283	Seconds per cubic foot
Grams (g)	0.0352	Ounces
Grams	0.0022	Pounds
Kilograms (kg)	2.205	Pounds
Degrees Celsius (°C)	1.8 (°C) + 32	Degrees Fahrenheit
Joules (J)	0.2388	Calories

References

Alexander, G.R. 1979. Predators of fish in coldwater streams. In: Clepper, H., ed. Predator-prey systems in fisheries management. Washington, DC: Sport Fishing Institute: 153–170.

Alsop, D.H.; Wood, C.M. 1997. The interactive effects of feeding and exercise on oxygen consumption, swimming performance and protein usage in juvenile rainbow trout (*Oncorhynchus mykiss*). Journal of Experimental Biology. 200: 2337–2346.

Avery, E.L.; Hine, R.L., eds. 1985. Sexual maturity and fecundity of brown trout in central and northern Wisconsin streams. Tech. Bull. No. 154. Madison, WI: Wisconsin Department of Natural Resources. 12 p.

Baltz, D.M.; Moyle, P.B. 1984. Segregation by species and size classes of rainbow trout, *Salmo gairdneri*, and Sacramento sucker, *Catostomus occidentalis*, in three California streams. Environmental Biology of Fishes. 10: 101–110.

Baltz, D.M.; Vondracek, B.; Brown, L.R.; Moyle, P.B. 1987. Influence of temperature on microhabitat choice by fishes in a California stream. Transactions of the American Fisheries Society. 116: 12–20.

Barnthouse, L.W.; Klauda, R.J.; Vaughan, D.S., eds. 1988. Science, law, and Hudson River power plants: a case study in environmental impact assessment. Bethesda, MD: American Fisheries Society. 347 p.

Barrett, J.C.; Grossman, G.D.; Rosenfeld, J. 1992. Turbidity-induced changes in reactive distance of rainbow trout. Transactions of the American Fisheries Society. 121: 437–443.

Bart, J. 1995. Acceptance criteria for using individual-based models to make management decisions. Ecological Applications. 5: 411–420.

Bartell, S.M.; Breck, J.M.; Gardner, R.H.; Brenkert, A.L. 1986. Individual parameter perturbation and error analysis of fish bioenergetics models. Canadian Journal of Fisheries and Aquatic Sciences. 43: 160–168.

Beer, W.N. 1999. Comparison of mechanistic and empirical methods for modeling embryo and alevin development in Chinook salmon. North American Journal of Aquaculture. 61: 126–134.

Behnke, R.J. 1992. Native trout of western North America. Bethesda, MD: American Fisheries Society. 275 p.

Booker, D.J.; Dunbar, M.J.; Ibbotson, A. 2004. Predicting juvenile salmonid drift-feeding habitat quality using a three-dimensional hydraulic-bioenergetic model. Ecological Modelling. 177: 157–177.

Bovee, K.D.; Lamb, B.L.; Bartholow, J.M.; Stalnaker, C.B.; Taylor, J.; Henriksen, J. 1998. Stream habitat analysis using the instream flow incremental methodology. Information and Technology Report USGS/BRD-1998-0004. Fort Collins, CO: U.S. Geological Survey, Biological Resources Division. 130 p.

Bowen, M.D. 1996. Habitat selection and movement of a stream-resident salmonid in a regulated river and tests of four bioenergetic optimization models. Logan, UT: Utah State University. 141 p. Ph.D. dissertation.

Braaten, P.J.; Dey, P.D.; Annear, T.C. 1997. Development and evaluation of bioenergetic-based habitat suitability criteria for trout. Regulated Rivers: Research and Management. 13: 345–356.

Bradford, M.J.; Higgins, P.S. 2001. Habitat-, season-, and size-specific variation in diel activity patterns of juvenile Chinook salmon (*Oncorhynchus tshawytscha*) and steelhead trout (*Oncorhynchus mykiss*). Canadian Journal of Fisheries and Aquatic Sciences. 58: 365–374.

Brandt, S.B.; Hartman, K.J. 1993. Innovative approaches with bioenergetics models: future applications to fish ecology and management. Transactions of the American Fisheries Society. 122: 731–735.

Brown, H.W. 1974. Handbook of the effects of temperature on some North American fishes. Canton, OH: American Electric Power Service Corp. 524 p.

Brown, L.C.; Barnwell, T.O.J. 1987. The enhanced stream water quality models QUAL2E and QUAL2E-UNCAS: documentation and user manual. EPA/600/3-87/07. Athens, GA: U.S. Environmental Protection Agency, Environmental Research Laboratory. 92 p.

Brown, L.R.; Moyle, P.B. 1991. Changes in habitat and microhabitat partitioning within an assemblage of stream fishes in response to predation by Sacramento squawfish (*Ptychocheilus grandis*). Canadian Journal of Fisheries and Aquatic Sciences. 48: 849–856.

Bull, C.D.; Metcalfe, N.B.; Mangel, M. 1996. Seasonal matching of foraging to anticipated energy requirements in anorexic juvenile salmon. Proceedings of the Royal Society of London B. 263: 13–18.

Butcher, G.; Parrish, M. 2006. Hydraulic simulations for regional fish modeling. Arcata, CA: Humboldt State University. 127 p. M.S. thesis.

Butler, P.J.; Day, N.; Namba, K. 1992. Interactive effects of seasonal temperature and low pH on resting oxygen uptake and swimming performance of adult brown trout *Salmo trutta*. Journal of Experimental Biology. 165: 195–212.

Carlander, K.D. 1969. Handbook of freshwater fishery biology. 3rd ed. Ames, IA: The Iowa State University Press. 752 p.

Clark, C.W.; Mangel, M. 2000. Dynamic state variable models in ecology. New York: Oxford University Press.

Cunjak, R.A.; Prowse, T.D.; Parrish, D.L. 1998. Atlantic salmon (*Salmo salar*) in winter: "The season of parr discontent"? Canadian Journal of Fisheries and Aquatic Sciences. 55(Suppl. 1): 161–180.

Cunningham, P.M. 2007. A sensitivity analysis of an individual-based trout model. Arcata, CA: Humboldt State University. 80 p. M.S. thesis.

Cutts, C.J.; Brembs, B.; Metcalfe, N.B.; Taylor, A.C. 1999. Prior residence, territory quality and life-history strategies in juvenile Atlantic salmon (*Salmo salar* L.). Journal of Fish Biology. 55: 784–794.

DeRobertis, A.; Ryer, C.H.; Veloza, A.; Brodeur, R.D. 2003. Differential effects of turbidity on prey consumption of piscivorous and planktivorous fish. Canadian Journal of Fisheries and Aquatic Sciences. 60: 1517–1526.

DeVries, P. 1997. Riverine salmonid egg burial depths: review of published data and implications for scour studies. Canadian Journal of Fisheries and Aquatic Sciences. 54: 1685–1698.

Diana, J.S.; Hudson, J.P.; Richard, J.; Clark, D. 2004. Movement patterns of large brown trout in the mainstream Au Sable River, Michigan. Transactions of the American Fisheries Society. 133: 34–44.

Dickerson, B.R.; Vinyard, G.L. 1999. Effects of high chronic temperatures and diel temperature cycles on the survival and growth of Lahontan cutthroat trout. Transactions of the American Fisheries Society. 128: 516–521.

Dill, L.M.; Ydenberg, R.C.; Fraser, A.H.G. 1981. Food abundance and territory size in juvenile coho salmon (*Oncorhynchus kisutch*). Canadian Journal of Zoology. 59: 1801–1809.

Drechsler, M. **2000.** A model-based decision aid for species protection under uncertainty. Biological Conservation. 94: 23–30.

Drechsler, M.; Frank, K.; Hanski, I.; O'Hara, R.B.; Wissel, C. **2003.** Ranking metapopulation extinction risk: from patterns in data to conservation management decisions. Ecological Applications. 13: 990–998.

Elliott, J.M. **1982.** The effects of temperature and ration size on growth and energetics of salmonids in captivity. Comparative Biochemistry and Physiology. 73: 81–91.

Elliott, J.M. **1994.** Quantitative ecology and the brown trout. New York: Oxford University Press. 286 p.

Elliott, J.M.; Hurley, M.A. **2000.** Daily energy intake and growth of piscivorous brown trout, *Salmo trutta*. Freshwater Biology. 44: 237–245.

Essington, T.E.; Quinn, T.P.; Ewert, V.E. **2000.** Intra- and inter-specific competition and the reproductive success of sympatric Pacific salmon. Canadian Journal of Fisheries and Aquatic Sciences. 57: 205–213.

Essington, T.E.; Sorensen, P.W.; Paron, D.G. **1998.** High rate of redd super-imposition by brook trout (*Salvelinus fontinalis*) and brown trout (*Salmo trutta*) in a Minnesota stream cannot be explained by habitat availability alone. Canadian Journal of Fisheries and Aquatic Sciences. 55: 2310–2316.

Fausch, K.D. **1984.** Profitable stream positions for salmonids: relating specific growth rate to net energy gain. Canadian Journal of Zoology. 62: 441–451.

Fausch, K.D.; Nakano, S.; Kitano, S. **1997.** Experimentally induced foraging mode shift by sympatric charrs in a Japanese mountain stream. Behavioral Ecology 8: 414–420.

Fraser, N.; Metcalfe, N.B. **1997.** The costs of becoming nocturnal: feeding efficiency in relation to light intensity in juvenile Atlantic salmon. Functional Ecology. 11: 385–391.

Fraser, N.; Metcalfe, N.B.; Thorpe, J.E. **1993.** Temperature-dependent switch between diurnal and nocturnal foraging in salmon. Proceedings of the Royal Society of London B. 252: 135–139.

From, J.; Rasmussen, G. **1984.** A growth model, gastric evacuation, and body composition in rainbow trout, *Salmo gairdneri* Richardson, 1836. Dana. 3: 61–139.

Gard, M. 1997. Techniques for adjusting spawning depth habitat utilization curves for availability. Rivers. 6: 94–102.

Gowan, C.; Fausch, K.D. 1996. Mobile brook trout in two high-elevation Colorado streams: re-evaluating the concept of restricted movement. Canadian Journal of Fisheries and Aquatic Sciences. 53: 1370–1381.

Gowan, C.; Fausch, K.D. 2002. Why do foraging stream salmonids move during summer? Environmental Biology of Fishes. 64: 139–153.

Grant, J.W.A.; Kramer, D.L. 1990. Territory size as a predictor of the upper limit to population density of juvenile salmonids in streams. Canadian Journal of Fisheries and Aquatic Sciences. 47: 1724–1737.

Gregory, R.S.; Levings, C.D. 1999. Turbidity reduces predation on migrating juvenile Pacific salmon. Transactions of the American Fisheries Society. 127: 275–285.

Griffiths, J.S.; Alderdice, D.F. 1972. Effects of acclimation and acute temperature experience on the swimming speed of juvenile coho salmon. Journal of the Fisheries Research Board of Canada. 29: 251–264.

Grimm, V.; Berger, U.; Bastiansen, F. [et al.]. 2006. A standard protocol for describing individual-based and agent-based models. Ecological Modelling. 198: 115–126.

Grimm, V.; Railsback, S.F. 2005. Individual-based modeling and ecology. Princeton, NJ: Princeton University Press. 428 p.

Groot, C.; Margolis, L., eds. 1991. Pacific salmon life histories. Vancouver, BC: UBC Press. 564 p.

Grossman, G.D.; Rincon, P.A.; Farr, M.D.; Ratajczak, R.E.J. 2002. A new optimal foraging model predicts habitat use by drift-feeding stream minnows. Ecology of Freshwater Fish. 11: 2–10.

Hanson, P.; Johnson, T.; Kitchell, J.; Schindler, D.E. 1997. Fish Bioenergetics 3.0. Madison, WI: University of Wisconsin Sea Grant Institute. 78 p.

Harvey, B.C. 1998. Influence of large woody debris on retention, immigration, and growth of coastal cutthroat trout (*Oncorhynchus clarki clarki*) in stream pools. Canadian Journal of Fisheries and Aquatic Sciences. 55: 1902–1908.

Harvey, B.C.; Nakamoto, R.J.; White, J.L. 1999. Influence of large woody debris and a bankfull flood on movement of adult resident coastal cutthroat trout (*Oncorhynchus clarki*) during fall and winter. Canadian Journal of Fisheries and Aquatic Sciences. 56: 2161–2166.

Harvey, B.C.; Railsback. S.F. 2007. Estimating multi-factor cumulative watershed effects on fish populations with an individual-based model. Fisheries. 32: 292–298.

Harvey, B.C.; Railsback, S.F. 2009. Exploring the persistence of stream-dwelling trout populations under alternative real-world turbidity regimes with an individual-based model. Transactions of the American Fisheries Society. 138: 348–360.

Harvey, B.C.; Stewart, A.J. 1991. Fish size and habitat depth relationships in headwater streams. Oecologia. 87: 336–342.

Harvey, B.C.; White, J.L. 2008. Use of benthic prey by salmonids under turbid conditions in a laboratory stream. Transactions of the American Fisheries Society. 137: 1756–1763.

Haschenburger, J.K. 1999. A probability model of scour and fill depths in gravel-bed channels. Water Resources Research. 35: 2857–2869.

Hawkins, D.K.; Quinn, T.P. 1996. Critical swimming velocity and associated morphology of juvenile coastal cutthroat trout (*Oncorhynchus clarki clarki*), steelhead trout (*Oncorhynchus mykiss*), and their hybrids. Canadian Journal of Fisheries and Aquatic Sciences. 53: 1487–1496.

Hayes, J.W.; Stark, J.D.; Shearer, K.A. 2000. Development and test of a whole-lifetime foraging and bioenergetics growth model for drift-feeding brown trout. Transactions of the American Fisheries Society. 129: 315–332.

Healey, M.C. 1991. Life history of Chinook salmon (*Oncorhynchus tshawytscha*). In: Groot, C.; Margolis, L., eds. Pacific salmon life histories. Vancouver, BC: UBC Press: 311–393.

Hendry, A.P.; Morbey, Y.E.; Berg, O.K.; Wenburg, J.K. 2003. Adaptive variation in senescence: reproductive lifespan in a wild salmon population. Proceedings of the Royal Society of London B. 271: 259–266.

Hilborn, R. 1997. Statistical hypothesis testing and decision theory in fisheries science. Fisheries. 22(10): 19–20.

Hilborn, R.; Mangel, M. 1997. The ecological detective. Princeton, NJ: Princeton University Press. 315 p.

Hill, J.; Grossman, G.D. 1993. An energetic model of microhabitat use for rainbow trout and rosyside dace. Ecology. 74: 685–698.

Hodgens, L.S.; Blumenshine, S.C.; Bednarz, J.C. 2004. Great blue heron predation on stocked rainbow trout in an Arkansas tailwater fishery. North American Journal of Fisheries Management. 24: 63–75.

Hokanson, K.E.F.; Kleiner, C.F.; Thorslund, T.W. 1977. Effects of constant temperatures and diel temperature fluctuations on specific growth and mortality rates and yield of juvenile rainbow trout (*Salmo gairdneri*). Journal of the Fisheries Research Board of Canada. 34: 639–648.

Holland, J.H. 1995. Hidden order: how adaptation builds complexity. Reading, MA: Perseus Books. 185 p.

Houston, A.I.; McNamara, J.M. 1999. Models of adaptive behavior: an approach based on state. Cambridge, England: Cambridge University Press. 378 p.

Hughes, N.F. 1992a. Ranking of feeding positions by drift-feeding arctic grayling (*Thymallus arcticus*) in dominance hierarchies. Canadian Journal of Fisheries and Aquatic Sciences. 49: 1994–1998.

Hughes, N.F. 1992b. Selection of positions by drift-feeding salmonids in dominance hierarchies: Model and test for arctic grayling (*Thymallus arcticus*) in subarctic mountain streams, interior Alaska. Canadian Journal of Fisheries and Aquatic Sciences. 49: 1999–2008.

Hughes, N.F.; Dill, L.M. 1990. Position choice by drift-feeding salmonids: model and test for arctic grayling (*Thymallus arcticus*) in subarctic mountain streams, interior Alaska. Canadian Journal of Fisheries and Aquatic Sciences. 47: 2039–2048.

Hughes, N.F.; Hayes, J.W.; Shearer, K.A.; Young, R.B. 2003. Testing a model of drift-feeding using 3-dimensional videography of wild brown trout in a New Zealand river. Canadian Journal of Fisheries and Aquatic Sciences. 60: 1462–1476.

Hurlbert, S.H. 1984. Pseudoreplication and the design of ecological field experiments. Ecological Monographs. 54: 187–211.

Hyvarinen, P.; Huusko, A. 2006. Diet of brown trout in relation to variation in abundance and size of pelagic fish prey. Journal of Fish Biology. 68: 87–98.

Inoue, M.; Nunokawa, M. 2002. Effects of longitudinal variations in stream habitat structure on fish abundance: an analysis based on subunit-scale habitat classification. Freshwater Biology. 47: 1594–1607.

Jager, H.I.; DeAngelis, D.L.; Sale, M.J.; Van Winkle, W.V.; Schmoyer, D.D.; Sabo, M.J.; Orth, D.J.; Lukas, J.A. 1993. An individual-based model for smallmouth bass reproduction and young-of-year dynamics in streams. Rivers. 4: 91–113.

Johnsson, J.I.; Nobbelin, F.; Bohlin, T. 1999. Territorial competition among wild brown trout fry: effects of ownership and body size. Journal of Fish Biology. 54: 469–472.

Jones, M.W.; Hutchings, J.A. 2002. Individual variation in Atlantic salmon fertilization success: implications for effective population size. Ecological Applications. 12: 184–193.

June, J.A. 1981. Life history and habitat utilization of cutthroat trout (*Salmo clarki*) in a headwater stream on the Olympic Peninsula, Washington. Seattle, WA: University of Washington. 112 p. M.S. thesis.

Keeley, E.R. 2001. Demographic responses to food and space competition by juvenile steelhead trout. Ecology. 82: 1247–1259.

Keeley, E.R.; Grant, J.W.A. 2001. Prey size of salmonid fishes in streams, lakes, and oceans. Canadian Journal of Fisheries and Aquatic Sciences. 58: 1122–1132.

Knapp, R.A.; Preisler, H.K. 1999. Is it possible to predict habitat use by spawning salmonids? A test using California golden trout (*Oncorhynchus mykiss aguabonita*). Canadian Journal of Fisheries and Aquatic Sciences. 56: 1576–1584.

Kondolf, G.M. 2000. Some suggested guidelines for geomorphic aspects of anadromous salmonid habitat restoration proposals. Restoration Ecology. 8: 48–56.

Lam, T.J. 1988. Environmental influences on gonadal activity in fish. In: Hoar, W.S.; Randall, D.J.; Donaldson, E.M., eds. Fish physiology, Volume IX, Reproduction, Part B: Behavior and fertility control. New York: Academic Press: 65–116.

Levin, S.A. 1999. Fragile dominion: complexity and the commons. Reading, MA: Helix Books. 256 p.

Lien, L. 1978. The energy budget of the brown trout population of Øvre Heimdalsvatn. Holarctic Ecology. 1: 279–300.

Lorek, H.; Sonnenschein, M. 1999. Modelling and simulation software to support individual-based ecological modelling. Ecological Modelling. 115: 199–216.

MacNutt, M.J.; Hinch, S.G.; Farrell, A.P.; Topp, S. 2004. The effect of temperature and acclimation period on repeat swimming performance in cutthroat trout. Journal of Fish Biology. 65: 342–353.

Magee, J.P.; McMahon, T.E.; Thurow, R.F. 1996. Spatial variation in spawning habitat of cutthroat trout in a sediment-rich stream basin. Transactions of the American Fisheries Society. 125: 768–779.

Magnusson, W.E. 2000. Error bars: are they the king's clothes? Bulletin of the Ecological Society of America. 81: 147–150.

Magoulick, D.D.; Wilzbach, M.A. 1999. Effect of temperature and macrohabitat on interspecific aggression, foraging success, and growth of brook trout and rainbow trout pairs in laboratory streams. Transactions of the American Fisheries Society. 127: 708–717.

Mangel, M.; Clark, C.W. 1986. Toward a unified foraging theory. Ecology. 67: 1127–1138.

Manly, B.F.J.; McDonald, L.L.; Thomas, D.L.; McDonald, T.L.; Erickson, W.P. 2002. Resource selection by animals: statistical design and analysis for field studies. 2nd ed. Boston, MA: Kluwer Academic Publishers. 221 p.

Metcalfe, N.B.; Fraser, N.H.C.; Burns, M.D. 1999. Food availability and the nocturnal vs. diurnal foraging trade-off in juvenile salmon. Journal of Animal Ecology. 68: 371–381.

Meyer, K.A.; Schill, D.J.; Elle, F.S.; Lamansky, J.A.J. 2003. Reproductive demographics and factors that influence length at sexual maturity of Yellowstone cutthroat trout in Idaho. Transactions of the American Fisheries Society. 132: 183–195.

Minar, N.; Burkhart, R.; Langton, C.; Askenazi, M. 1996. The Swarm simulation system: a toolkit for building multi-agent simulations. Working Paper 96-06-042. Santa Fe, NM: Santa Fe Institute. 11 p.

Morbey, Y.E.; Ydenberg, R.C. 2003. Timing games in the reproductive phenology of female Pacific salmon (*Oncorhynchus* spp.). American Naturalist. 161: 284–298.

Morin, A.; Dumont, P. 1994. A simple model to estimate growth rate of lotic insect larvae and its value for estimating population and community production. Journal of the North American Benthological Society. 13: 357–367.

Moyle, P.B.; Baltz, D.M. 1985. Microhabitat use by an assemblage of California stream fishes: developing criteria for instream flow determinations. Transactions of the American Fisheries Society. 114: 695–704.

Moyle, P.B.; Marchetti, M. 1992. Temperature requirements of rainbow trout and brown trout. Unpublished report. On file with: P. Moyle, Department of Wildlife, Fish, and Conservation Biology, University of California, Davis, CA 95616.

Myrick, C.A. 1998. Temperature, genetic, and ration effects on juvenile rainbow trout (*Oncorhynchus mykiss*) bioenergetics. Davis: University of California. 166 p. Ph.D. dissertation.

Myrick, C.A.; Cech, J.J., Jr. 2000. Temperature influences on California rainbow trout physiological performance. Fish Physiology and Biochemistry. 22: 245–254.

Myrick, C.A.; Cech, J.J., Jr. 2003. The physiological performance of golden trout at water temperatures of 10–19 °C. California Fish and Game. 89: 20–29.

Nelson, R.L.; Platts, W.S.; Casey, O. 1987. Evidence for variability in spawning behavior of interior cutthroat trout in response to environmental uncertainty. Great Basin Naturalist. 47: 480–487.

NeXT 1995. Object-oriented programming and the Objective-C language. Redwood City, CA: NeXT Computer Inc. http://www.gnustep.org/resources/ documentation/ObjectivCBook.pdf. (December 15, 2008)

Nielsen, J.L. 1992. Microhabitat-specific foraging behavior, diet, and growth of juvenile coho salmon. Transactions of the American Fisheries Society. 121: 617–634.

Nislow, K.; Folt, C.; Seandel, M. 1998. Food and foraging behavior in relation to microhabitat use and survival of age-0 Atlantic salmon. Canadian Journal of Fisheries and Aquatic Sciences. 55: 116–127.

Orth, D.J. 1987. Ecological considerations in the development and application of instream flow-habitat models. Regulated Rivers: Research & Management. 1: 171–181.

Pacific Gas and Electric Company [PG&E]. 1994. Evaluation of factors causing variability in habitat suitability criteria for Sierra Nevada trout. Report 009.4-94.5. San Ramon, CA: Department of Research and Development. 31 p.

Poff, N.L.; Huryn, A.D. 1998. Multi-scale determinants of secondary production in Atlantic salmon (*Salmo salar*) streams. Canadian Journal of Fisheries and Aquatic Sciences. 55: 201–217.

Post, J.R.; Parkinson, E.A.; Johnston, E.A. 1998. Spatial and temporal variation in risk to piscivory of age-0 rainbow trout: patterns and population level consequences. Transactions of the American Fisheries Society. 127: 932–942.

Power, M.E. 1987. Predator avoidance by grazing fishes in temperate and tropical streams: importance of stream depth and prey size. In: Kerfoot, W.C.; Sih, A., eds. Predation: direct and indirect impacts on aquatic communities. Hanover, NH: University Press of New England: 333–352.

Preall, R.J.; Ringler, N.H. 1989. Comparison of actual and potential growth rates of brown trout (*Salmo trutta*) in natural streams based on bioenergetic models. Canadian Journal of Fisheries and Aquatic Sciences. 46: 1067–1076.

Quinn, T.P.; Buck, G.B. 2001. Size- and sex-selective mortality of adult sockeye salmon: bears, gulls, and fish out of water. Transactions of the American Fisheries Society. 130: 995–1005.

Railsback, S.F. 1999a. Reducing uncertainties in instream flow studies. Fisheries 24(4): 24–26.

Railsback, S.F. 1999b. Tools for individual-based stream fish models. TR-114006. Palo Alto, CA: Electric Power Research Institute. 21 p.

Railsback, S.F. 2001. Concepts from complex adaptive systems as a framework for individual-based modelling. Ecological Modelling. 139: 47–62.

Railsback, S.F.; Cunningham, P.M.; Lamberson, R.H. [N.d.]. A strategy for parameter sensitivity and uncertainty analysis of individual-based models. Manuscript in preparation. On file with: S. Railsback, Lang, Railsback & Assoc., 250 California Ave., Arcata, CA 95521.

Railsback, S.; Harvey, B. 2001. Individual-based model formulation for cutthroat trout, Little Jones Creek, California. Gen. Tech. Rep. PSW-GTR-182. Albany, CA: U.S. Department of Agriculture Forest Service, Pacific Southwest Research Station. 80 p.

Railsback, S.F.; Harvey, B.C. 2002. Analysis of habitat selection rules using an individual-based model. Ecology. 83: 1817–1830.

Railsback, S.F.; Harvey, B.C.; Hayes, J.W.; LaGory, K.E. 2005. Tests of theory for diel variation in salmonid feeding activity and habitat use. Ecology. 86: 947–959.

Railsback, S.F.; Harvey, B.C.; Lamberson, R.H.; Lee, D.E.; Claasen, N.J.; Yoshihara, S. 2002. Population-level analysis and validation of an individual-based cutthroat trout model. Natural Resource Modeling. 15: 83–110.

Railsback, S.F.; Kadvany, J. 2008. Demonstration flow assessment: judgment and visual observation in instream flow studies. Fisheries. 33: 217–227.

Railsback, S.F.; Lamberson, R.H.; Harvey, B.C.; Duffy, W.E. 1999. Movement rules for spatially explicit individual-based models of stream fish. Ecological Modelling. 123: 73–89.

Railsback, S.F.; Rose, K.A. 1999. Bioenergetics modeling of stream trout growth: temperature and food consumption effects. Transactions of the American Fisheries Society. 128: 241–256.

Railsback, S.F.; Stauffer, H.B.; Harvey, B.C. 2003. What can habitat preference models tell us? Tests using a virtual trout population. Ecological Applications. 13: 1580–1594.

Rand, P.S.; Stewart, D.J.; Seelbach, P.W.; Jones, M.L.; Wedge, L.R. 1993. Modeling steelhead population energetics in Lakes Michigan and Ontario. Transactions of the American Fisheries Society. 122: 977–1001.

Reiser, D.W.; White, R.G. 1983. Effects of complete redd dewatering on salmonid egg-hatching success and development of juveniles. Transactions of the American Fisheries Society. 112: 532–540.

Ropella, G.E.P.; Railsback, S.F.; Jackson, S.K. 2002. Software engineering considerations for individual-based models. Natural Resource Modeling. 15: 5–22.

Saltelli A.; Tarantola, F.; Campolongo, F.; Ratto, M. 2004. Sensitivity analysis in practice: a guide to assessing scientific models. New York: Halsted Press. 232 p.

Schmidt, D.; O'Brien, W.J. 1982. Planktivorous feeding ecology of Arctic grayling (*Thymallus arcticus*). Canadian Journal of Fisheries and Aquatic Sciences. 39: 475–482.

Schneider, M.J.; Connors, T.J. 1982. Effects of elevated water temperature on the critical swimming speeds of yearling rainbow trout, *Salmo gairdneri*. Journal of Thermal Biology. 7: 227–229.

Simpkins, D.G.; Hubert, W.A.; Martinez Del Rio, C.; Rule, D.C. 2003a. Interacting effects of water temperature and swimming activity on body composition and mortality of fasted juvenile rainbow trout. Canadian Journal of Zoology. 81: 1641–1649.

Simpkins, D.G.; Hubert, W.A.; Martinez Del Rio, C.; Rule, D.C. 2003b. Physiological responses of juvenile rainbow trout to fasting and swimming activity: effect of body composition and condition indices. Transactions of the American Fisheries Society. 132: 576–589.

Stearley, R. F.1992. Historical ecology of Salmoninae, with special reference to *Oncorhynchus*. In: Mayden, R.L., ed. Systematics, historical ecology, and North American freshwater fishes. Stanford, CA: Stanford University Press: 622–658.

Swarm Development Group [SDG]. 2000. Document set for Swarm 2.1.1. Santa Fe, NM: Swarm Development Group. http://www.swarm.org/swarmdocs-2.2/set/set.html (December 15, 2008)

Sweka, J.A.; Hartman, K.J. 2001. Influence of turbidity on brook trout reactive distance and foraging success. Transactions of the American Fisheries Society. 130: 138–146.

Taylor, S.E.; Egginton, S.; Taylor, E.W. 1996. Seasonal temperature acclimatisation of rainbow trout: cardiovascular and morphometric influences on maximal sustainable exercise level. Journal of Experimental Biology. 199: 835–845.

Thomas R. Payne & Associates [TRPA]. 1998. User's Manual, RHABSIM 2.0. Arcata, CA.

Thorpe, J.E.; Mangel, M.; Metcalfe, N.B.; Huntingford, F.A. 1998. Modelling the proximate basis of salmonid life-history variation, with application to Atlantic salmon, *Salmo salar* L. Evolutionary Ecology. 12: 581–599.

Thut, R.N. 1970. Feeding habits of the dipper in southwestern Washington. Condor. 72: 234–235.

Underwood, T.J.; Bennett, D.H. 1992. Effects of fluctuating flows on the population dynamics of rainbow trout in the Spokane River of Idaho. Northwest Science. 66: 261–268.

Valdimarsson, S.K.; Metcalfe, N.B. **2001.** Is the level of aggression and dispersion in territorial fish dependent on light intensity? Animal Behavior. 61: 1143–1149.

Valdimarsson, S.K.; Metcalfe, N.B.; Thorpe, J.E.; Huntingford, F.A. **1997.** Seasonal changes in sheltering: effect of light and temperature on diel activity in juvenile salmon. Animal Behavior. 54: 1405–1412.

Van Winkle, W.; Jager, H.I.; Holcomb, B.D. **1996.** An individual-based instream flow model for coexisting populations of brown and rainbow trout. TR-106258. Palo Alto, CA: Electric Power Research Institute. 105 p.

Van Winkle, W.; Jager, H.I.; Railsback, S.F.; Holcomb, B.D.; Studley, T.K.; Baldrige, J.E. **1998.** Individual-based model of sympatric populations of brown and rainbow trout for instream flow assessment: model description and calibration. Ecological Modelling. 110: 175–207.

Vinyard, G.L.; Winzeler, A. **2000.** Lahontan cutthroat trout (*Oncorhynchus clarki henshawi*) spawning and downstream migration of juveniles into Summit Lake, Nevada. Western North American Naturalist. 60: 333–341.

Vogel, J.L.; Beauchamp, D.A. **1999.** Effects of light, prey size, and turbidity on reaction distances of lake trout (*Salvelinus namaycush*) to salmonid prey. Canadian Journal of Fisheries and Aquatic Sciences. 56: 1293–1297.

Volpe, J.P.; Anholt, B.R.; Glickman, B.W. **2000.** Competition among juvenile Atlantic salmon (*Salmo salar*) and steelhead (*Oncorhynchus mykiss*): relevance to invasion potential in British Columbia. Canadian Journal of Fisheries and Aquatic Sciences. 58: 197–207.

Waldrop, M.M. **1992.** Complexity: the emerging science at the edge of order and chaos. New York: Simon and Schuster. 380 p.

White, J.L.; Harvey, B.C. **2007.** Winter feeding success of stream trout under different streamflow and turbidity conditions. Transactions of the American Fisheries Society. 136: 1187–1192.

Wilcock, P.R.; Barta, A.F.; Shea, C.C.; Kondolf, G.M.; Matthews, W.V.G.; Pitlick, J. **1996.** Observations of flow and sediment entrainment on a large gravel-bed river. Water Resources Research. 32: 2897–2909.

Williams, J.G. **1996.** Lost in space: minimum confidence intervals for idealized PHABSIM studies. Transactions of the American Fisheries Society. 125: 458–465.

Wilzbach, M.A. 1985. Relative roles of food abundance and cover in determining the habitat distribution of stream-dwelling cutthroat trout (*Salmo clarki*). Canadian Journal of Fisheries and Aquatic Sciences. 42: 1668–1672.

Xenopoulos, M.A.; Lodge, D.M.; Alcamo, J.; Marker, M.; Schulze, K.; VanVuuren, D.P. 2005. Scenarios of freshwater fish extinctions from climate change and water withdrawal. Global Change Biology. 11: 1557–1564.

Yalin, M.S. 1977. Mechanics of sediment transport. 2nd ed. Oxford, United Kingdom: Permagon Press. 298 p.

Young, K.A. 2003. Evolution of fighting behavior under asymmetric competition: an experimental test with juveniles salmonids. Behavioral Ecology. 14: 127–134.

Ziemer, R.R.; Rice, R.M.; Lisle, T.E. 1991. Modeling the cumulative watershed effects of forest management strategies. Journal of Environmental Quality. 20: 36–42.

Index of Parameters and Variables

This publication is available online at http://www.fs.fed.us/psw/. You may also order additional copies of it by sending your mailing information in label form through one of the following means. Please specify the publication title and series number.

Fort Collins Service Center

Web site	http://www.fs.fed.us/psw/
Telephone	(970) 498-1392
FAX	(970) 498-1122
E-mail	rschneider@fs.fed.us
Mailing address	Publications Distribution Rocky Mountain Research Station 240 West Prospect Road Fort Collins, CO 80526-2098

Pacific Southwest Research Station
800 Buchanan Street
Albany, CA 94710

Federal Recycling Program
Printed on Recycled Paper